Occurrence of *Legionella* in Groundwater

The mission of the Awwa Research Foundation (AwwaRF) is to advance the science of water to improve the quality of life. Funded primarily through annual subscription payments from over 1,000 utilities, consulting firms, and manufacturers in North America and abroad, AwwaRF sponsors research on all aspects of drinking water, including supply and resources, treatment, monitoring and analysis, distribution, management, and health effects.

From its headquarters in Denver, Colorado, the AwwaRF staff directs and supports the efforts of over 700 volunteers, who are the heart of the research program. These volunteers, serving on various boards and committees, use their expertise to select and monitor research studies to benefit the entire drinking water community.

Research findings are disseminated through a number of technology transfer activities, including research reports, conferences, videotape summaries, and periodicals.

Occurrence of *Legionella* in Groundwater

Prepared by:
Serge Riffard, **Susan Springthorpe**, **Leo Filion**, and **Syed A. Sattar**
Centre for Research on Environmental Microbiology (CREM)
University of Ottawa, Ottawa, ON, Canada

With the participation of:
Teresa Brooks and **Raegan Osicki**
Centre for Research on Environmental Microbiology (CREM)
University of Ottawa, Ottawa, ON, Canada

Marilyn Lee
Ryerson University, Toronto, ON, Canada

and

David Abrial
Laboratoire de Biologie Animale et Appliquée, Faculté des Sciences,
Université de Saint Etienne, France

Sponsored by:
Awwa Research Foundation
6666 West Quincy Avenue, Denver, CO 80235-3098

Published by:

Awwa Research Foundation
Advancing the Science of Water®

IWA Publishing

DISCLAIMER

This study was funded by the Awwa Research Foundation (AwwaRF). AwwaRF assumes no responsibility for the content of the research study reported in this publication or for the opinions or statements of fact expressed in the report. The mention of trade names for commercial products does not represent or imply the approval or endorsement of AwwaRF. This report is presented solely for informational purposes.

Published by IWA Publishing, Alliance House, 12 Caxton Street, London SW1H 0QS, UK

Telephone: +44 (0) 20 7654 5500; Fax: +44 (0) 20 7654 5555; Email: publications@iwap.co.uk;
Web: **www.iwapublishing.com**

AwwaRF report number 90985F. Originally published by AwwaRF for its subscribers in 2003.
IWA Publishing version published 2004.

ISBN 1 84339 8818

CONTENTS

TABLES

FIGURES

FOREWORD

The Awwa Research Foundation is a nonprofit corporation that is dedicated to the implementation of a research effort to help utilities respond to regulatory requirements and traditional high-priority concerns of the industry. The research agenda is developed through a process of consultation with subscribers and drinking water professionals. Under the umbrella of a Strategic Research Plan, the Research Advisory Council prioritizes the suggested project based upon current and future needs, applicability, and past work; the recommendations are forwarded to the Board of Trustees for final selection. The foundation also sponsors research projects through an unsolicited proposal process; the Collaborative Research, Research Applications, and Tailored Collaboration programs; and various joint research efforts with organizations such as the U.S. Environmental Protection Agency, the U.S. Bureau of Reclamation, and the Association of California Water Agencies.

This publication is a result of one of these sponsored studies, and it is hoped that its findings will be applied in communities throughout the world. The following report serves not only as a means of communicating the results of the water industry's centralized research program but also as a tool to enlist the further support of the nonmember utilities and individuals.

Projects are managed closely from their inception to the final report by the foundation's staff and large cadre of volunteers who willingly contribute their time and expertise. The foundation serves a planning and management function and awards contracts to other institutions such as water utilities, universities, and engineering firms. The funding for this research effort comes primarily from the Subscription Program, through which water utilities subscribe to the research program and make an annual payment proportionate to the volume of water they deliver and consultants and manufacturers subscribe based on their annual billings. The program offers a cost-effective and fair method for funding research in the public interest.

A broad spectrum of water supply issues is addressed by the foundation's research agenda: resources, treatment and operations, distribution and storage, water quality and analysis, toxicology, economics, and management. The ultimate purpose of the coordinated effort is to assist water suppliers to provide the highest possible quality of water economically and reliably. The true benefits are realized when the results are implemented at the utility level. The foundation's trustees are pleased to offer this publication as a contribution toward that end.

This is the first comprehensive study on the occurrence of *Legionella* in groundwater. Members of this species are emerging pathogens. While they have been known to be present in surface waters, this investigation has clearly demonstrated that they are widely distributed in groundwater sources as well. We hope that the methods and results reported here will form the basis for further studies on protecting groundwater-based supplies from such opportunistic pathogens and in a better assessment of any human health risks that may be associated with the use of groundwaters.

Edmund G. Archuleta, P.E.
Chair, Board of Trustees
Awwa Research Foundation

James F. Manwaring, P.E.
Executive Director
Awwa Research Foundation

ACKNOWLEDGMENTS

We wish to thank all utilities that contributed to this study. We are especially grateful to the staff of "Water Utility A" for their excellent cooperation. In particular, we would like to underline the efforts from Ms. M. Pryor ("Water Utility A" Laboratory Manager) for all the data and also the work provided to coordinate sampling events. We also acknowledge the help of the American Water Works Service Company for providing us samples and coordinating sampling events. We thank Dr. Mark LeChevallier, Ms. Stacey Dulaney and Ms. Cheryl Norton in particular for their participation in the project.

We are also very grateful to our scientific collaborators mentioned on the title page: Prof. Marilyn Lee (School of Occupational & Public Health, Ryerson University, Toronto, Canada) for her help in examining the water and biofilm samples for protozoa, Dr. David Abrial (Faculté des Sciences, Université de Saint Etienne, France) for the statistical analyses of the data, and Steve Douglas, Teresa Brooks and Raegan Osicki for their excellent and thorough technical work.

We thank Dr. Frances Jamieson and Mr. Patrick Tang (Central Public Health Laboratory, Toronto, Canada) for providing us with *Legionella* strains.

Suggestions and guidance from Mr. Emerson Lomaquahu, Ms. Jennifer Warner (Awwa Research Foundation Project Managers) and members of the Project Advisory Committee (Dr. David Metge, Dr. Pat Rusin, and Dr. Christian Volk) are also gratefully acknowledged.

We also thank Mr. Gilles Morier and members of our University's Research and Financial services for their help in dealing with all contractual aspects of this study.

EXECUTIVE SUMMARY

Members of the genus *Legionella* are ubiquitous in surface waters, but we know much less about their occurrence in groundwaters, particularly those that are not known to be under the influence of surface waters. This is important since wells are the source of potable water for more than 106 million Americans and Canadians, and many others around the world. This study was therefore initiated to understand better the occurrence and ecology of legionellae in groundwaters by analyzing water and biofilm samples from several locations in the United States and Canada. Additionally, ecological data were used, where available, to evaluate if biotic or abiotic factors could act as indicators for the presence of *Legionella* spp.

In Phase I of this investigation, water and biofilm samples obtained from only two sites (warm-water utilities in Florida) were analyzed for legionellae. In Phase II, while continuing monthly samples from the first Florida site, several other warm- and cold-water sites from across the U.S. and Canada were also surveyed. The U.S. sites were from the states of Arizona, Florida, Illinois, Michigan, Missouri, Nebraska, New Jersey and Texas; samples in Canada came from New Brunswick and Ontario. Most of the water utilities enrolled in the study had groundwater systems that were not known to be under the direct influence of surface water.

Since this study aimed at looking at the water source rather than the distribution system, water and biofilm samples were mainly taken at the well head before entering the treatment plant and distribution system. Depending on the resources available at a given utility, different strategies were used for the collection of water and biofilm specimens. Whenever possible, long-established biofilms were collected to ensure a better view of *Legionella* colonization of a given system. Because legionellae may also act as pioneer bacteria, biofilm colonization of pipe sections was investigated when long-established biofilms (such as in casings) could not be collected. *Legionella* were detected by combining a quantitative culture method (considered the gold standard) with a qualitative semi-nested PCR assay. The use of immuno-magnetic separation (IMS) was also evaluated to selectively capture legionellae from some of the samples.

In Phase I, the overall feasibility of the experimental approach was demonstrated using relatively few samples taken from production wells at warm water participating utilities. In Phase II, a total of 105 samples (48 water and 57 biofilm) were cultured for legionellae. Of these, 23 (48%) water and 13 (23%) biofilm samples were *Legionella*-positive, giving an overall percentage of positive samples of 34% by cultivation. When combining results from the two study phases, 56% of water samples (37/66) and 29% of biofilm samples (20/69) were found positive by cultivation. PCR was able to detect legionellae in 32 of the 105 (30%) samples tested in Phase II; PCR inhibitors proved to be a problem, especially in biofilm samples. The use of a commercial DNA extraction kit appeared to improve *Legionella* detection in Phase II. Based on the wells sampled, rather than individual samples, 11 of 12 (92%) wells were positive for legionellae in Phase I and 27 of 34 (79%) sampled were positive for legionellae in Phase II. Utilities with only negative samples were 0/2 in Phase I and 2/16 in Phase II. Each of these two had submitted only single samples from one well. Thus it is possible that repeated sampling or inclusion of other wells would have changed the picture. Although most of the wells were surveyed only once in this study, two wells from one utility in Phase II were resampled on a monthly basis for a year. In these samples it became clear that there was great temporal variability in the numbers and types of legionellae detected. It is impossible to say whether this applies to all wells that are not heavily contaminated with a single strain but it seems likely.

IMS, which was used in Phase II only, was tested in combination with cultivation to reduce interference from faster-growing contaminants and in combination with PCR to try and resolve the problems with PCR inhibitors. The IMS technique was evolving as it was being optimized during the course of the study. Relatively few samples were processed under the optimal conditions established, and the technique is not yet mature enough to use routinely. However, IMS can simplify sample processing and could potentially be used by water utilities for detecting legionellae in their own treatment and distribution systems. It showed some promise but clearly needs more development work and validation before it could be used routinely on field samples.

The results from this study show that:

a. several species of *Legionella* could be readily recovered from warm or cold groundwater systems not known to be under the direct influence of surface water, even though the numbers and types of the legionellae recovered showed considerable variation;
b. within the one water utility sampled regularly, temporal variations were observed in the types and numbers of legionellae recovered;
c. none of the water quality and ecological parameters available showed any definitive correlation with the presence or absence of *Legionella* in the samples tested, but the possible association with the presence or absence of certain metals deserves further investigation;
d. limited analysis of samples positive for *Legionella* showed them to also contain several types of protozoa, possibly associated with *Legionella* survival and/or growth, although experimental evidence of this is not provided by this study.

The findings of this first comprehensive study on the occurrence of *Legionella* showed them to be widely distributed in groundwaters from a variety of sites in the U.S. and Canada. Whereas it is impossible to determine the point(s) of initial entry of legionellae into underground aquifers, it is possible to speculate that their contamination with these opportunistic pathogens could be the result of unknown influences of surface water, and or preferential flow channels accessed from the surface during recharge. Some of the wells tested are, however, quite deep and depending on local hydrogeology, surface influences may be relatively distant from the site of the production wells.

Although the numbers of legionellae recovered in this study were relatively low in most cases, the efficiency of recovery is unknown and may also be low. It is presumed that in the absence of treatment such levels would be sufficient to seed a distribution system where subsequent amplification could occur. Given their relative resistance to disinfection, widespread *Legionella* colonization of distribution systems may present a health risk, particularly to those who are debilitated or immunocompromised. Legionellae may become dominant organisms in distribution system biofilms, and clumps of them dislodged from biofilms and then aerosolized may present the greatest risk to exposed individuals. The importance of *Legionella* ecology and their relationship with protozoa, especially entrapment and survival in protozoan cycts, needs to be recognized in order to understand their behaviour and distribution in the environment. Such associations, and how they affect the fluctuating numbers and types of legionellae available, is an important issue to be considered in quantitative risk assessment for this group of bacteria.

CHAPTER 1
BACKGROUND AND RATIONALE

INTRODUCTION

Microorganisms pathogenic for humans can be released into the subsurface and hence to groundwater in many different ways. Septic fields, deteriorating infrastructure and sprayed or spread liquid or solid municipal and animal wastes are among the most common sources of subsurface contamination. In addition, known or unsuspected linkage(s) of surface water to groundwater can influence the quality and microbial content of groundwater. Legionellae are well-known contaminants of surface water but very little has been ascertained regarding their presence and distribution in groundwater.

Groundwater in North America

Overall, more than 100 million Americans get their drinking water from underground sources (Yates, 1997); in rural areas 95% of potable water is estimated to come from wells (Bitton and Gerba, 1984). In fact, reliance on groundwater resources is increasing further due to population growth as well as deterioration in surface water quality and its diversion for competing uses such as agriculture. In some areas depletion of underground aquifers is significant, and dropping water tables can also be associated with declining quality because of increasing salinity and other reasons. It is estimated by the U.S. Environmental Protection Agency (EPA) that 72% of the public water supply systems in America that use groundwater do not disinfect. Overall, 18 million people are served untreated groundwater from community water systems (CWSs), 4 million are served untreated groundwater from non-transient non-community water systems (NTNCWSs) and an additional 13 million are served untreated groundwater from transient non-community water systems (TNCWSs) (Figure 1-1) meaning that 35 million people may be receiving untreated water.

In Canada, over six million people, or 20% of the population, rely on groundwater for domestic use. Approximately two thirds, or four million, of these users live in rural settings. In many areas, wells produce more reliable and less expensive water supplies than those obtained from nearby lakes, rivers and streams. The remaining two million users are located primarily in smaller municipalities where groundwater provides the primary source for their water supply systems. For instance, 100% of Prince Edward Island's population and over 60% of the population of New Brunswick and the Yukon Territory rely on groundwater to meet their domestic needs. Furthermore, the predominant use of groundwater varies by province. In Ontario, Prince Edward Island, New Brunswick, and the Yukon, the largest users of groundwater are municipalities; in Alberta, Saskatchewan, and Manitoba, the agricultural industry for livestock watering; in British Columbia, Quebec and the Northwest Territories, industry; and in Newfoundland and Nova Scotia, rural domestic use. Prince Edward Island is almost totally dependent on groundwater for all its need.

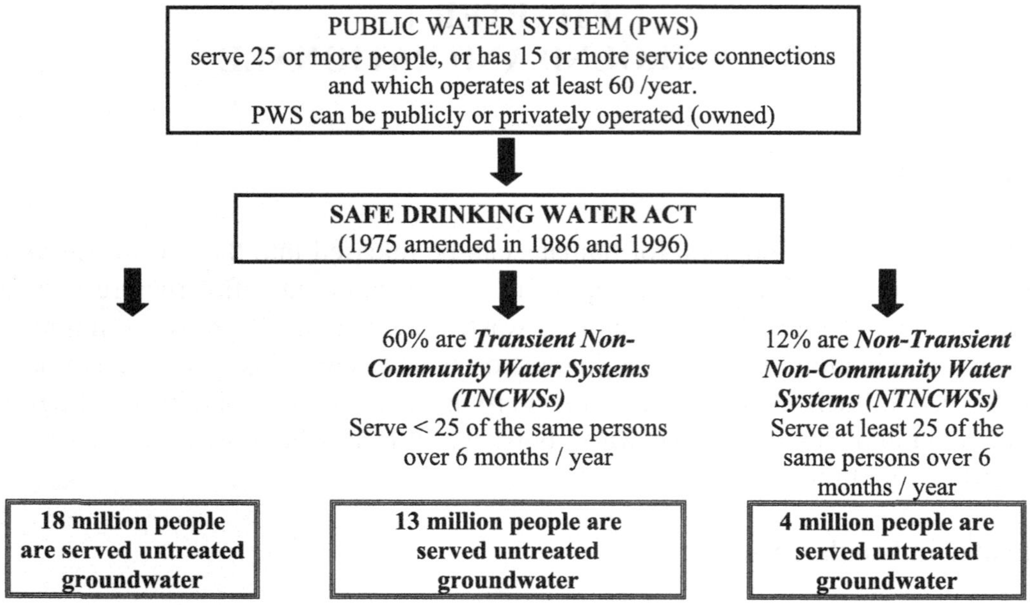

Figure 1.1 U.S. population relying on untreated groundwater for its domestic use—summarized from the U.S. EPA proposed Groundwater Rule

Groundwater Contamination

Groundwater systems are artificially classified depending on whether the aquifer is known to be under the influence of surface water or not. When groundwaters are under the influence of surface water, contaminated surface source(s) can play a significant role in their contamination. Prior to distribution, water from such sources is required to be treated using a combination of filtration and disinfection, or comparable technologies. Even if an aquifer is not known to be under the direct influence of surface water, if it is unconfined, it clearly must have some links with the surface for recharge to occur. These unconfined aquifers can become contaminated during recharge through seepage of chemical and/or microbial pollutants from the soil through preferential flow channels. How well microbial contaminants can be retained by soils during such seepage is governed by a complex interplay of local soil characteristics, climatic and weather conditions, as well as the distance between the site of contamination and the water table. Local geological formations and previous construction activities can also influence how readily groundwater contamination from surface or near surface sources will occur. Unseen linkage(s) through fractured bedrock or karst geological features can carry contaminants rapidly over considerable distances, horizontally and, occasionally, vertically. Karst is generally associated with large underground caverns and streams, but much smaller fissures and irregularities can be responsible for rapid recharge and carriage of contaminants into the aquifer. In spite of potential contamination hazards, groundwater supplies which are not known to be under the direct influence of surface waters, may receive no treatment at all, or only disinfection.

LEGIONELLA

Historical Data

The medical significance of the genus *Legionella* was first recognized after an outbreak of pneumonia afflicting members of the Pennsylvania American Legion during a meeting in Philadelphia in 1976 (Winn, 1988; Winn, 1995). This outbreak led to the recognition of the clinical entity Legionnaires' disease (LD). Subsequent to the isolation and identification of *L. pneumophila* as the etiologic agent of LD, it was discovered that outbreaks had occurred as long ago as 1957 and sporadic cases even earlier. The question of airborne dissemination of these bacteria was first raised after an epidemic of respiratory disease at St. Elizabeth's hospital (Washington, D.C., USA), in 1965. Another significant episode was the acute respiratory disease that afflicted employees and visitors in a health department in Pontiac (Michigan) suggesting a second milder form of LD (Edelstein and Meyer, 1994), now called Pontiac Fever. Members of the genus *Legionella* are thus a cause of both community-acquired and nosocomial pneumonia. They account for about 1-4% of the pneumonia necessitating hospitalization (Breiman, 1993).

It is estimated that 8,000 to 18,000 cases of LD occur in the U.S. annually and 10 to 20% of these are related to outbreaks; however, these numbers may vary according to which sources are considered (some estimates suggest that 25,000 to 100,000 cases of LD occur in the U.S. each year). Recent findings related to the discovery of not-yet-cultivable *Legionella* species, formerly named *Legionella*-Like Amoebal Pathogens (LLAPs) (Fry et al. 1991a; Benson et al. 1995; Butler et al. 1995; Adeleke, 1996; Hookey et al. 1996; Newsome et al. 1998; Adeleke et al. 2001), suggest that these bacteria are probably an underestimated cause of pneumonia. Incomplete reporting from many countries suggests that the actual numbers of *Legionella* infections may be much higher on a global basis (Infuso, Hubert, and Etienne, 1998). This view is supported by the wide variations in the reported incidence of such infections in European countries where the numbers may vary from 0.1 to 30 cases/million (Joseph and Hutchinson, 1996).

Most outbreaks, as well as most of the sporadic cases investigated, have been linked to either tap water or cooling towers (Stout et al. 1992; Bhopal, 1995). It has been calculated that if at least half of all *Legionella* infections were acquired from tap water, this would represent at least 10,000 cases per year. Considering an average mortality of 25%, this would represent approximately 25,000 deaths per year (Kool, Carpenter, and Fields, 1999).

Taxonomic Data

Since the first description of the family Legionellaceae in 1979 (Brenner, Steigerwalt, and McDade, 1979), 48 species have been described (Lo Presti et al. 2001) (Table 1.1). This includes *L. lytica*, a species not-yet culturable on BCYE media (Benson et al. 1996; Hookey et al. 1996). Other LLAPs have recently been described (Adeleke et al. 2001). *L. pneumophila* remains the most frequently isolated species from human patients although eighteen other species, including *L. lytica*, have also been recovered from human cases of pneumonia (Hookey et al. 1996; Lo Presti et al. 1997). Some species, such as *L. parisiensis* (Brenner et al. 1985; Lo Presti et al. 1997), *L. gormanii* (Morris et al. 1980; Griffith et al. 1988; Towns, Fisher, and Moore, 1994) and *L. jordanis* (Cherry et al. 1982; Thacker et al. 1988; Baty et al. 1997) were first isolated from environmental specimens before being recognized as human pathogens. Even though *L. pneumophila* still remains the most common cause of LD (Bartlett, Macrae, and MacFarlane,

Table 1.1
Described *Legionella* species and ability to cause human disease

Species	Sero-groups	Type strains	Source (ATCC)	Species implicated in human diseases (Y = yes; N = no)
L. pneumophila	1	Philadelphia-1	33152	Y (Brenner, Steigerwalt, and McDade, 1979)
	2	Togus-1	33154	
	3	Bloomington-2	33155	Y (McKinney et al. 1979)
	4	Los Angeles-1	33156	Y (Watts et al. 1980)
	5	Dallas-1E	33216	Y (Brenner et al. 1988)
	5	Micu-B	33735	Y (England et al. 1980)
	6	Chicago-2	33215	N (Brenner et al. 1988)
	7	Chicago-8	33823	Y (McKinney et al. 1980)
	8	Concord-3	35096	N (Bibb et al. 1983)
	9	IN-23-G1-C2	35289	Y (Bissett, Lee, and Lindquist, 1983)
	10	Leiden-1	43283	Y (Edelstein et al. 1984)
	11	797-PA-H	43130	Y (Meenhorst et al. 1985)
	12	570-CO-H	43290	Y (Thacker et al. 1986)
	13	82 A 3105	43736	Y (Thacker et al. 1987)
	14	1169-MN-H	43073	Y (Lindquist et al. 1988)
	"15"	Lansing 3	35251	Y (Benson et al. 1988)
				Y (Brenner et al. 1988)
L. adelaidensis		1762-AUS-E	49625	N (Benson et al. 1991)
L. anisa		WA-316-C3	35292	Y (Thacker et al. 1990)
L. beliardensis		Montbéliard A1	700512	N (Lo Presti et al. 2001)
L. birminghamensis		1407-AL-H	43702	Y (Wilkinson et al. 1987)
L. bozemanii	1	WIGA	33217	Y (Brenner et al. 1980)
L. bozemanii	2	Toronto 3	35545	Y (Tang et al. 1984)
L. brunensis		441-1	43878	N (Wilkinson et al. 1988)
L. cherrii		ORW	35252	Y (Brenner et al. 1985; Fang, Yu, and Vickers, 1989)
L. cincinnatiensis		72-OH-H	43753	Y (Thacker et al. 1988)
L. drozanskii		LLAP-1	700990	N* (Adeleke et al. 2001)
L. dumoffii		NO-23	33279	Y (Brenner et al. 1980)
L. erythra	1	SE-32A-C8	35303	N (Brenner et al. 1985)
L. erythra	2	LC 217		Y (Saunders, Doshi, and Harrsion, 1992)

(continued)

Table 1.1 (Continued)

Species	Sero-groups	Type strains	Source (ATCC)	Species implicated in human diseases (Y = yes; N = no)
L. fallonii		LLAP-10	700992	N* (Adeleke et al. 2001)
L. fairfieldensis		1725-AUS-E	49588	N (Thacker et al. 1991)
L. feeleii	1	WO-44C	35072	Y (Herwaldt et al. 1984)
L. feeleii	2	691-WI-H	35849	Y (Thacker et al. 1985; Pietsch, Graeber, and Lorenz, 1995)
L. geestiana		1308	49504	N (Dennis et al. 1993)
L. genomospecies	1	2055-AUS-E	51913	N (Benson et al. 1996)
L. gormanii		LS-13	33297	Y (Brenner et al. 1985; Griffith et al. 1988)
L. gratiana		LYON 8420412	49413	N (Bornstein et al. 1989)
L. gresilensis		Gréoux 11D13	700509	N (Lo Presti et al. 2001)
L. hackeliae	1	Lansing 2	35250	Y (Brenner et al. 1985)
L. hackeliae	2	798-PA-H	35999	Y (Wilkinson et al. 1985)
L. israelensis		Bercovier-4	43119	N (Bercovier et al. 1986)
L. jamestowniensis		JA-26-G1-E2	35298	N (Brenner et al. 1985)
L. jordanis		BL-540	33623	Y (Thacker et al. 1988)
L. lansingensis		1677-MI-H	49751	Y (Thacker et al. 1992)
L. londiniensis		1477	49505	N (Dennis et al. 1993)
L. londiniensis	2	Mulhouse B26	700510	N (Lo Presti et al. 1996)
L. longbeachae	1	Long Beach 4	33462	Y (McKinney et al. 1981)
L. longbeachae	2	Tucker 1	33484	Y (Bibb et al. 1981)
L. lytica		L2	PCM 2298	N (Drozanski, 1991)
L. maceachernii		PX-1-G2-E2	35300	Y (Brenner et al. 1985)
L. micdadei		Tatlock	33218	Y (Hebert, Steigerwalt, and Brenner, 1980; O'Connell, Bangsborg, and Cianciotto, 1995)
L. moravica		316-36	43877	N (Wilkinson et al. 1988)
L. nautarum		1224	49506	N (Dennis et al. 1993)
L. oakridgensis		OR-10	33761	Y (Orrison et al. 1983; Tang, Toma, and MacMillan, 1985; Lo Presti et al. 2000)

(continued)

Table 1.1 (Continued)

Species	Sero-groups	Type strains	Source (ATCC)	Species implicated in human diseases (Y = yes; N = no)
L. parisiensis		PF-209C-C2	35299	Y (Lo Presti et al. 1997)
L. quateirensis		1335	49507	N (Dennis et al. 1993)
L. quinlivanii	1	43830	43830	N (Benson et al. 1989)
L. quinlivanii	2	LC870	NCTC12434	N (Birtles et al. 1991)
L. rowbothamii		LLAP-6	700991	N* (Adeleke et al. 2001)
L. rubrilucens		WA-270A-C2	35304	N (Brenner et al. 1985)
L. sainthelensi	1	Mt St Helens-4	35248	Y (Campbell et al. 1984; Benson et al. 1990)
L. sainthelensi	2	1489-CA-H	49322	Y (Benson et al. 1990)
L. santicrucis		SC-63-C7	35301	N (Brenner et al. 1985)
L. shakespearei		214	49655	N (Verma et al. 1992)
L. spiritensis		Mt St Helens-9	35249	N (Brenner *et al.* 1985)
L. steigerwaltii		SC-18-C9	35302	N (Brenner et al. 1985)
L. taurinensis		Turin I N°1	700508	N (Lo Presti et al. 1996)
L. tucsonensis		1087-AZ-H	49180	Y (Thacker et al. 1989)
L. wadsworthii		81-716A	33877	Y (Edelstein et al. 1982)
L. waltersii		2074-AUS-E	51914	N (Benson et al. 1996)
L. worsleiensis		1347	49508	N (Dennis et al. 1993)

* Serologic evidence associating this species with respiratory disease in humans has been reported.
ATCC, American Type Culture Collection.
NCTC, National Collection Type Culture.
PCM, Polish Culture Collection.

1986; Fang et al. 1990; Bornstein and Fleurette, 1992; Edelstein and Meyer, 1994; Harrington et al. 1996), non-*pneumophila* species are increasingly being incriminated (Fang et al. 1989), especially in cases of LD in the immunocompromised (Harrington et al. 1996). Moreover, it has been reported that infections with multiple *Legionella* species (Dowling et al. 1983; Dowling et al. 1984; Tang and Krishnan, 1993), multiple *Legionella* serogroups (Tang and Krishnan, 1993; Jalaludin et al. 1997; Lück, Wenchel, and Helbig, 1998) or even multiple *Legionella* strains can occur (Horbach, Naumann, and Fehrenbach, 1988; Winn, 1988).

Clinical Considerations

There are two diseases associated with legionellae; i.e. LD and Pontiac Fever. Both diseases can sometimes be caused by the same species. The conditions under which an individual develops either LD or Pontiac Fever are not fully understood but may depend on the health status of the individual, the degree of exposure to the organism, or the strain-specific virulence.

Pontiac Fever is a self-limiting, febrile illness usually affecting healthy individuals. Because it is self-limiting, this form of *Legionella* infection is not as much of a concern and remarkably little is known about its prevalence. On the other hand, the symptoms of LD range from a mild cough and low-grade fever to fevers in excess of 40°C accompanying rapidly progressive pneumonia with multi-organ failure. The case fatality rate is usually 20 to 40% during an outbreak (Winn, 1988; Edelstein, 1993; Edelstein and Meyer, 1994; Fields, 1997). Although statistics for LD are well documented, the incidence of Pontiac Fever in the general community is unknown.

LD can be acquired by inhalation of aerosols or microaspiration of water containing legionellae (Bollin et al. 1985). The smallest dose needed to cause infection is believed to be about 10 CFU/mL. Aerosols with a droplet size of about 5 μm can readily enter the respiratory tract, colonize the lung, and induce disease (Berendt et al. 1980; Meenhorst et al. 1983). Human-to-human transmission of LD has not been recorded. Also, drinking of contaminated water is not known to be a risk factor. However, the widespread use of showers for bathing can pose a potential risk for users.

Inhalation of particles of soil containing legionellae can also cause infection. *Legionella longbeachae*, for example, has been incriminated in cases of LD after exposure to potting soil in Australia (McKinney et al. 1981; Steele, Moore, and Sangster, 1990), Japan (Koide et al. 1999) and in the United States.

Ecology

Known Reservoirs

Bacteria of the family Legionellaceae are common inhabitants of both natural and anthropogenic aquatic habitats (Fliermans et al. 1981; Stout, Yu, and Best, 1985; Stout, Yu, and Muraca, 1985; Fliermans, 1996). Waters often serve as the source of *Legionella* spp. Aerosols from cooling towers, for example, are frequently implicated in outbreaks of LD (Bhopal and Barr, 1990; Bentham and Broadbent, 1993; Shelton, Flanders, and Morris, 1994; Castellani-Pastoris et al. 1997). Although it is not definitively known if Legionellaceae are able to be free living, several lines of evidence suggest that in natural environments, legionellae grow exclusively within other organisms. However, a free-living state cannot be completely excluded since *Legionella* can find, in particular ecosystems such as biofilms, ecological conditions favorable to their replication. Colonies of *L. pneumophila* have been seen growing independently in biofilms, suggesting that these organisms are capable of replication in the water environment outside the cells of hosts such as free-living protozoa (Rogers and Keevil, 1992).

Abiotic Influences

Temperature. Water temperature is generally regarded as an important limiting environmental factor for the growth of *Legionella;* it has been suggested (Yee and Wadowsky, 1982) that *Legionella* spp. fail to multiply in aquatic environments below 25°C. However, these organisms appear to have adapted to temperature ranges much wider than those for bacteria normally thought to be of human health significance. *Legionella* have been detected in waters with temperatures range from 5.7 to 63°C (Fliermans, 1984).

Ecological studies demonstrated that *L. pneumophila* could multiply in water over a temperature range of 25°C (Yee and Wadowsky, 1982) to 42-50°C (Wadowsky et al. 1985; Wilkinson et al. 1988; Winn, 1988; Zacheus and Martikainen, 1996) and could be frequently associated with biotopes where temperatures can rise to 60°C (Fliermans, 1983). However, some studies showed that when temperatures rise to 50°C, the viability of the bacteria may be reduced (Dennis, Green, and Jones, 1984; Stout, Best, and Yu, 1986). Other *Legionella* species (*L. bozemanii, L. micdadei*) show a variable resistance to high temperatures (Hernandez, Delattre, and Ogre, 1983; Dennis, Green, and Jones, 1984). On the other hand, legionellae are able to survive at low temperatures since they have been successfully isolated from frozen rivers (Fliermans, 1983) as well as contaminated ice and water dispensed from ice machines (Stout, Yu, and Muraca, 1985; Bangsborg et al. 1995; Graman, Quilan, and Rank, 1997). Even if about 37% of the utilities having groundwater under the influence of surface water have water temperatures of 15°C or higher (data available from Awwa Research Foundation WATERSTATS database), there are many groundwater sources with temperatures consistently below 10°C. Under these conditions, it is tempting to speculate that *Legionella* are not able to multiply. However, their natural reservoirs in water may be inside protozoa and accumulation in biofilms; thus, not only survival but also multiplication may be possible at these colder temperatures. Adaptation to varying temperatures is well known in many species of bacteria and the range of temperatures tolerated by any particular legionellae may be influenced by the temperatures in which they are accustomed to surviving and potentially multiplying; in that sense the adaptation and replication of the host protozoa may also be critical.

Groundwater temperatures normally reflect mean annual air temperature and may range from a few degrees above freezing in parts of North America to 30°C or more in the tropics. It is also notable that *Legionella* are considered to be recovered more often from environmental samples or clinical specimens at 35°C even if cultivation at 30°C (or less) might be more consistent with their true physiological state in nature (Rogers and Keevil, 1992; Rogers et al. 1994a; Rogers et al. 1994b). An early study on the occurrence of *Legionella* and other aquatic bacteria in chemically contaminated groundwater treated by aeration failed to detect *Legionella* by culture in both water and scraped biomass (Spino, Rice, and Geldreich, 1984). Although the lower temperatures characteristic of the groundwaters studied (<12°C) may have been a major hindrance in the detection of this organism, cultivation methods may also have been inadequate.

A review on groundwater temperatures was conducted from the WATERSTATS database on representative utilities that have and use groundwater resources. Numerous utilities (27 out of 72) with groundwater wells under the influence of surface water have waters that reach 15°C or more. These utilities located across in the United States (632 wells in 29 states) might possess numerous wells of interest for the present study. Data kindly provided by the Laboratory Manager (participating Water utility "A," Florida) showed that their well water temperature is constant, ranging from 24 to 27°C, with wells situated between 300 and 900 feet deep. Roughly, on the basis of data provided on 13 sites, associated heterotrophic microflora counts increased from June to October, whereas total coliform and *E. coli* are rarely detected through the survey, suggesting that these wells are not under the immediate influence of surface water. Another survey (personal communication) showed that among 13 states, 88 sites have groundwater temperatures above 15°C. These utilities, located in Alabama, Florida, Indiana, Texas, and Washington, have groundwater with temperatures in the range of 20 to 30°C.

Finally, water temperature has been found to be an important rate-controlling factor in regrowth of coliforms (LeChevallier, Schulz, and Lee, 1991). It is also likely to be important for the growth of many other bacteria which are potential prey for the protozoan predators that act as hosts for legionellae.

Chemical compounds. Substances frequently reported in municipal drinking water systems which enhance *Legionella* growth, are total organic carbon, zinc and copper (States et al. 1987). The effect of copper is controversial since available data on biofilm formation and growth of *L. pneumophila* (in particular) in potable water systems explain that surfaces of copper are poorly colonized by cultures of *L. pneumophila* (Schofield and Locci, 1985). Among metals, iron (ferrous iron is found generally in waters with low level of dissolved oxygen) and manganese are those most commonly found in groundwaters. These metals are important in the metabolism of numerous bacteria and protozoa (Marshall et al. 1997), and can contribute to slime production. It is important to remember that most *Legionella* species require iron either for their extracellular growth (Goldoni et al. 1991; Dumais-Pope, O'Connell, and Cianciotto, 1996) or for intracellular growth in host cells (Reeves et al. 1981; Byrd and Horwitz, 1987; Byrd and Horwitz, 1989). It has been shown that numerous iron salts (ferric nitrate, ferrous sulfate, ammoniacal ferric sulfate, ferric chloride, ferric phosphate, hematin and hemin) may be needed to support in vitro growth of *L. pneumophila* (Feeley et al. 1978; Pine et al. 1979; Reeves et al. 1981; Johnson et al. 1982; O'Connell, Hickey, and Cianciotto, 1996). A decrease in iron concentration in vitro is responsible for morphological variations in *L. pneumophila* (Nowicki et al. 1987), and a decrease in virulence and intracellular replication in particular (Gebran et al. 1994; James et al. 1995). No specific chemical factor has yet been described as predictive of the presence or absence of legionellae in the environment (Fliermans, 1984).

Other environmental factors. Other environmental factors need to be considered—dissolved oxygen content (normally very low in groundwater), redox potentials (from +500 to –200 mV), and total organic carbon (TOC). Most commonly, aquifers are considered to be electron donor limited. Many aquifer systems in this category are composed of sediments or rocks tens to hundreds of million years old from which metabolizable organic carbon has been progressively removed over geologic time. The other category of aquifers consists of those that are electron acceptor limited. In these systems, there is an excess of electron donor available (usually organic carbon) such that microbial metabolism is constrained by the availability of electron acceptors. These aquifers have often been contaminated by human activities, and are more likely to be under the influence of surface waters. In surface water systems (or groundwaters under the direct influence of surface waters), most coliform occurrences have been associated with TOC levels greater than 2.4 mg/L and AOC levels greater than 50 µg/L (LeChevallier, Schulz, and Lee, 1991). It has also been shown that peaks in AOC level and rainfall were positively correlated to coliform occurrence (LeChevallier, Schulz, and Lee, 1991).

Rainfall may have a great impact on nutrient loading and consequently on the occurrence of bacteria and protozoa in groundwaters (Atherholt et al. 1998). Thus, monitoring weather events during dry season(s), especially where groundwater is under the influence of surface water, might need to be considered. Indeed, small amounts of rainfall following a long dry spell may have a greater impact on nutrient loading than a large amount of rainfall in the wet season (LeChevallier, Schulz, and Lee, 1991). A combination of parameters may also be important: environmental (rainfall, temperature, etc.), geological and hydrological (wells depth, unsaturated soil around wells, etc.), nutritional (AOC and TOC, etc.), and chemical parameters could be related to occurrence of Legionella in groundwater in the same way as that of coliform bacteria in groundwater (Abbaszadegan et al. 1998) or water distribution systems.

Biotic Influences

Groundwater systems probably have a higher degree of microbial diversity than is usually recognized, although some local oligotrophic conditions may not be favorable to survival of many microbial pathogens. Ecosystems that are electron acceptor-limited possess trophic structure and exhibit materials cycling. It is likely then, as a part of the hydrological cycle, that some of the groundwater wells may contain many of the microorganisms encountered in soil. Flagellates, amoebae, photosynthetic microorganisms, and fungi, are also likely to be represented in these groundwaters. Among these groundwater reservoirs, protozoan hosts as well as bacterial species synergistic for *Legionella* growth are likely to be encountered.

Knowing the above information, this study also used available relevant quantitative data on heterotrophic microflora and, in particular, on the presence of potential indicators such as the coliform group where this was provided by the participating utilities. A limited number of the samples were also analyzed for the presence of protozoa that may serve as natural amplifiers for legionellae. It is now recognized that specific interactions may exist between some *Legionella* species and some protozoan hosts. Moreover, some *Legionella* species or strains might be better able to colonize protozoa than others suggesting that particular clones might be more widely distributed, or better able to persist in the environment, than others. Such an "ecological advantage" might be responsible for the predominance of certain clones and could explain why, sometimes, most of the clinical cases in a particular geographical area seem to be correlated with a particular epidemiological isolate (Forey et al. 1998).

Bacteria. Survival of legionellae in water is highly dependent on presence of cyanobacteria (Berendt, 1981; Bohach and Snyder, 1983b), algae (Tison and Seidler, 1983; Hume and Hann, 1984; Tison, 1987) and other non-*Legionella* microorganisms that might stimulate (Stout et al. 1986) or inhibit their growth (Paszko-Kolva et al. 1993a). It is accepted that some bacteria of the heterotrophic flora might also supply most of the necessary nutrients (Wadowsky and Yee, 1983; Fliermans, 1984; Stout et al. 1985a; Wadowsky and Yee, 1985; Stout et al. 1986b; Gomez-Lus et al. 1993; Paszko-Kolva et al. 1993a). Environmental bacteria may also inhibit growth of *Legionella* spp. in water. Inhibitory effects of heterotrophic bacteria such as *Pseudomonas aeruginosa*, or *Aeromonas hydrophila* on the cultivation of *L. dumoffii* have been shown (Paszko-Kolva et al. 1993a). The mechanism(s) by which bacteria might compete with or antagonize *Legionella* spp. in water environments or associated biofilms remain unclear. However, bacteriocinogenic strains are likely to exert an antagonistic activity towards *Legionella* spp. (Gomez-Lus et al. 1993), and this mechanism might be the one implicated when there is a close physical association, for instance, in biofilms.

Protozoa. *Legionella* can invade and multiply within the cells of a large variety of protozoa and be incorporated into amoebic cysts (Rowbotham, 1980; Fields et al. 1984; Fields et al. 1989; Rowbotham, 1989; Fields, 1993; Fields et al. 1993; Steele, 1993; Steele and McLennan, 1996). This last-mentioned fact is very important because a variety of free-living amoebae and ciliated protozoa have been isolated along with *Legionella* spp. from waters (Nahapetian et al. 1991; Smith-Somerville et al. 1991; Wadowsky et al. 1991) and legionellae multiply intracellularly within amoebae much as they do within human macrophages. Moreover, as well as providing protection and supporting proliferation in hostile environments, amoebae may also contribute to the environmental persistence of Legionellaceae by inducing a poly-hydroxybutyrate (endogenous source of carbon) -rich phenotype, which is physiologically more suited for extracellular survival in low-nutrient environments (James et al. 1999). Finally, a specific

Table 1.2
Protozoa known to support the growth of legionellae

Type	Species
Amoebae	*Acanthamoeba castellanii*
	A. polyphaga
	A. palestiniensis
	A. royreba
	A. culbertsonii
	Naegleria gruberi
	N. fowleri
	N. lovaniensis
	N. jadini
	Hartmannella vermiformis
	H. cantabrigiensis
	Vahlkampfia jugosa
	Willaertia magna
	Echinamoeba exudans
	Dictyostelium discoideum
Ciliated protozoa	*Tetrahymena pyriformis*
	T. vorax

Note: The protozoa listed in Table 1.2 do not support all *Legionella* species. Also, strains within a given species may also vary with regards to their association with protozoa.

protozoan, *Hartmannella vermiformis*, has been frequently identified in several outbreaks of LD (Segal and Shuman, 1999). The list of protozoan species supporting *Legionella* is increasing (Table 1.2), and recently it has even been reported (Molmeret et al. 2000) that legionellae can be amplified even further in a second protozoan following phagocytosis of the first host. However, the host ranges of *Legionella* species are generally unknown since systemic studies of potential hosts have not been carried out.

As mentioned above, *L. pneumophila* is a parasite of at least 13 species of amoebae and ciliated protozoa (Fields, 1996). *Legionella*-contaminated hospital hot water systems have been shown to be frequently colonized by amoebae identified as *H. vermiformis*, *Echinamoebae* spp., *Saccamoebae* spp., and *Vahlkampfia* spp. Other amoebae such as *Acanthamoeba*, *Naegleria* or *Vanella* have also been demonstrated by sampling swabs (Rohr et al. 1998). Finally, *Tetrahymena* spp. and *Cyclidium* spp. have also been isolated from cooling tower water containing *L. pneumophila*, a site responsible for an LD outbreak (Barbaree et al. 1986).

L. pneumophila-infected amoebae are infectious particles *in vivo* (Brieland et al. 1996; Brieland et al. 1997). This fact supports the hypothesis that inhaled protozoa may serve as cofactors in the pathogenesis of pulmonary disease (Brieland et al. 1997). *L. pneumophila* virulence for amoebae is required for maximal intrapulmonary growth of the bacteria in mice co-inoculated with *H. vermiformis*. Inhaled amoebae may potentiate intrapulmonary growth of *L. pneumophila* by providing a niche for bacterial replication (Brieland et al. 1997). Moreover, *A. castellanii*-grown *L. pneumophila* have been shown to be at least 10-fold more invasive for macrophages than were *L. pneumophila* cells grown on agar (Cirillo, Falkow, and Tompkins, 1994).

Two *Acanthamoeba* species have been shown to expel vesicles containing living *L. pneumophila* cells. Vesicles ranged from 2.1 to 6.4 μm in diameter and theoretically could contain several hundred bacteria (Berk et al. 1998). According to Rowbotham (1986), a 5-μm-diameter vesicle (of respirable size) might contain a range of 365 to 1,483 bacteria for a vesicle that is 90% full.

If amoebae really support survival and growth of legionellae in hot water systems, they probably belong to hartmannellae and other thermoresistant forms. *H. vermiformis*, the most common protozoan isolated from potable water in the U.S., is frequently found in waters implicated as the source of the bacteria in outbreaks of LD. A growth-supporting effect of *H. vermiformis* on legionellae in tap water has been described (Wadowsky et al. 1991). A correlation has been demonstrated between the occurrence of *H. vermiformis* and *L. pneumophila* (serogroup 1) in the water systems of hospitals (Sanden et al. 1992).

Some species of *Naegleria* interact with *Legionella* spp. and may help in the dissemination of *Legionella* in water (Ma et al. 1990). Significant numbers of *Naegleria* spp. have been found in water layers containing filamentous cyanobacteria and eubacteria, which serve as food sources. *Naegleria* spp. are found to increase in numbers in waters contaminated with coliforms. A large number of organisms have been isolated from waters with increased iron and manganese concentrations (Marshall et al. 1997).

Acanthamoeba spp. pervade the environment and can be found in tap water, fresh water, and soil. *Acanthamoeba* spp. are also known to be a replicative niche for *Legionella* spp. (Berk et al. 1998). *L. pneumophila* survive in *Tetrahymena vorax*, a hymenostome-ciliated protozoan that prefers lower temperatures. *L. pneumophila* did not multiply in the ciliate when incubated at 20 to 22°C but vacuoles containing *L. pneumophila* were retained in the cells for a substantially longer time than vacuoles with *E. coli* (Smith-Somerville et al. 1991). Thus, *T. vorax* may act as a reservoir for *Legionella* in cold water-related environments.

Whenever possible, we also tested the samples for the presence of protozoa that may serve as natural amplifiers for *Legionella*. This was done with the expectation that the presence of certain types of protozoa may explain the predominance of specific species of *Legionella* (Forey et al. 1998). Indeed, some *Legionella* spp. may multiply only in specific protozoan hosts as demonstrated for *L. anisa* with *H. vermiformis* (Fields et al. 1990).

Biofilms

The colonization of plumbing systems by *L. pneumophila* is well documented (Rogers et al. 1991) and the growth of the organism in potable water supplies may lead to human infection. Detection of low numbers of Legionellaceae in aquatic environments is frequently observed (Lye et al. 1997). Those *Legionella* cells are likely to be planktonic, present in protozoa or recently shed from biofilms. It is accepted that adherent microbial consortia or biofilms occur within the water system and contain numbers of *Legionella* that are higher than those in the water phase (Rogers et al. 1994a; Rogers et al. 1994b). Therefore, examination of this sessile phase might not necessitate additional concentration steps in order to detect *Legionella* in such environments. If *Legionella* cells are present in groundwater, they might be concentrated in biofilms especially if those biofilms are recovered from devices such as water meters or from inserted coupons or even pipes.

It is not known whether the inoculum size (i.e., the number of *Legionella* in the water source) will influence colonization success of a potential pre-existing biofilm. *Legionella* cells could also act as pioneer bacteria in the establishment of a biofilm and may persist in this biofilm. Once again, it is not known how *Legionella* cells take part, through ecological succession, in biofilm initiation or persistence.

Once established in biofilms, little is known about *Legionella* interactions with other biofilm microorganisms. However, it is possible that larger *Legionella* populations might result from mutualistic interaction between bacteria. Because they are microaerophiles, legionellae also may find, in a biofilm niche, a suitable low-oxygen environment for enhanced growth (Mauchline and Keevil, 1991). Based on a chemostat model of a water distribution system, inoculated with a complex microbial consortium containing an indigenous population of *L. pneumophila* with associated bacteria and protozoa, legionellae are thought to comprise only a low proportion (about 2%) of biofilm flora (Rogers and Keevil, 1992). *L. pneumophila*, in particular, is present in tight micro-colonies among the biofilm flora suggesting that the organism is actively growing within the consortium and that growth of the legionellae could be sustained in the absence of host organisms (Rogers and Keevil, 1992). Recent studies in a distribution system suggest, however, that under certain conditions *Legionella* can be the predominant biofilm bacterium (M. Pryor – personal communication).

Another recent study on biofilms found *L. pneumophila* in the majority of the samples tested. Legionellaceae were detected in biofilms from rivers or brooks, lakes or ponds, and fountains having temperatures of 3.0 to 22.5°C; pH value, 7.3 to 8.7; conductivity, 300 to 1050 µS/cm (Grimm et al. 1998). The results obtained in this study are in good agreement with earlier distribution studies of *L. pneumophila* and *Legionella* spp. Once again, it seems that Legionellaceae are capable of surviving wide ranges of physical and biological conditions; and that there is no obvious "preference" of the bacteria for a certain habitat (Grimm et al. 1998).

Current State of Knowledge on *Legionella* in Groundwater

Very little is known about the presence of *Legionella* in groundwaters. One study (Lye et al. 1997) reported 93% of groundwater samples to be *Legionella*-positive by PCR, while only 7% of them were culture-positive; *L. pneumophila* was not detected. The lack of agreement between the results of PCR and cultivation was ascribed to the possibility that most *Legionella* cells in the samples were in a non-cultivable state. Subsequently, however, the PCR detection kit (EnviroAmp™ *Legionella* kit, Perkin-Elmer) used in this investigation was found to be unreliable (Brown et al. 1997), and was withdrawn from the market. Lye et al. (1997) did not include any information on the wells sampled, or the physico-chemical characteristics of the water examined. They also analyzed a limited number of biofilm samples and this may have also affected the low recovery rate by cultivation. Since other investigators also failed to detect *L. pneumophila* using the EnviroAmp™ *Legionella* kit (Maiwald et al. 1995; Okpara et al. 1996), it is tempting to speculate that the 5S rRNA target gene is more efficiently amplified by PCR than the *mip* target gene because the 5S rRNA gene is present in multiple copies on the genome. Initially, Bej et al. (1990) demonstrated that multiplex PCR based upon adjusting the ratios of primers for *Legionella* 5S rRNA and *mip* amplification was unsuccessful, whereas initiating *mip* amplification for a few cycles prior to multiplex PCR worked (Bej et al. 1990). Other possibilities include samples under the method detection threshold, or that some sub-inhibitory substances in the samples could have affected the amplification yield of the *mip* gene.

The EnviroAmp™ kit was initially developed for the detection of *Legionella* in water samples (Palmer et al. 1993; Atlas, William, and Huntington, 1995; Palmer et al. 1995; Ng et al. 1997), but it was later used with clinical specimens as well (Kessler et al. 1993; Matsiota-Bernard et al. 1994; Ramirez et al. 1996). However, it could detect only 25 of the 43 *Legionella* species described, thus also undermining its application in testing groundwater samples. In spite of further modifications (Brown et al. 1997; Okpara et al. 1996; Palmer et al. 1993; Tompkins and Loutit, 1993; Lye et al. 1997; Miyamoto et al. 1997; Ng et al. 1997), the kit and PCR assays in general have remained limited in their application for the study of environmental samples.

Some reports suggest that groundwaters may have been implicated in the spread of Legionnaires' disease (LD). In Arizona (Kioski et al. 1997), legionellae were recovered from a private well supplying water to a hospital with cases of LD. However, epidemiological data found no correlation with contamination from the suspected well. *L. pneumophila* serogroup 1 was isolated from the well and *L. pneumophila* serogroup 6 was found to infect patients and to colonize taps and showers in patients' rooms. Phenotypic variation amongst genotypically homogeneous *L. pneumophila* serogroup 1 isolates has been shown (Harrison et al. 1990). Whether this explanation may be relevant in this context is unknown. Additional examples of LD attributed to groundwater include inhalation of *Legionella* in water emerging from a well under high pressure (Miragliotta et al. 1992) and in the Savoie region of France, legionellae have been isolated from spring waters at spa therapy centers (Bornstein, Marmet, and Surgot, 1988; Bornstein et al. 1989). In France, regular surveys of water from thermal spas have led to the isolation of numerous *Legionella* species (Bornstein, Marmet, and Surgot, 1988; Bornstein, Marmet, and Surgot, 1989). Prospective surveys have shown that deep wells were uncontaminated (unpublished data) while those closer to the surface were contaminated with as many as 15 *Legionella* species, providing evidence for contamination of this type of groundwater by percolated surface water or soil. Spring water temperature was typically 39-45°C and contained mainly sulfates and thiosulfates in the form of calcium and sodium salts (Bornstein et al. 1989). It is not known whether sulfates or thiosulfates are more favorable for the growth of *Legionella* than other chemical substances. No correlations were found with the heterotrophic microflora recovered (unpublished data).

More recently, legionellae have been isolated from water distribution systems of apartment buildings that received their potable water as a blend of surface and groundwaters (Zacheus and Martikainen, 1994). Legionellae appeared more often and with higher concentrations in hot water systems using water processed in surface water plants than in water systems associated with groundwater plants (Zacheus and Martikainen, 1994).

Current Regulations/Policies on *Legionella* in Water

Microbial standards for drinking water restrict the numbers of particular organisms allowed to be present. Legionellae are primarily known to be associated with surface water and their presence is regulated through the Surface Water Treatment Rule (SWTR) promulgated by the U.S. EPA. The SWTR also applies to groundwater sources known to be under the direct influence of surface water. Thus and presently, only surface water systems and systems using groundwater under the direct influence of surface water are required to disinfect their water supplies. EPA will propose, for groundwater sources that are not under the direct influence of surface water and to any system that mixes surface and groundwater (if the groundwater is added directly to the distribution system without prior treatment), a Groundwater Rule (GWR) that will include a Groundwater Disinfection Rule intended to set up maximum contaminant level goals (MCLGs) and minimum levels of disinfection for numerous pathogens. Under this GWR, a utility may not be required to apply any disinfectant if it can show that its groundwater source is not vulnerable to fecal contamination. This study has been partly undertaken to evaluate whether an association between fecal contaminants and *Legionella* can be found in groundwater. However, it is also necessary to mention that legionellae have never been found associated with fecal indicators in the past. Currently, EPA is seeking additional evidence for the occurrence of particular pathogens including *Legionella*, *Pseudomonas aeruginosa* and *Mycobacterium avium-intracellulare* in groundwater before considering them for regulations.

Disinfection

Ozonation, silver-copper ionization, UV radiation, elevated temperatures, and periodic hyperchlorination have all been investigated for their effectiveness in eliminating, or controlling, legionellae (specifically *L. pneumophila*) from water systems. In addition, design changes and incorporation of materials less hospitable to biofilm formation have also been tried to reduce the risk of colonization of water distribution system with legionellae.

Light and UV Radiation

Ultraviolet (UV) radiation (wavelength of 260-280 nm) has been demonstrated as an effective method for prevention of *Legionella* colonization of water fixtures under the following conditions (Makin and Hart, 1993): the UV radiation was continuous, the lamp(s) were situated near the point of use, and UV was combined with filtration (Lin et al. 1992). Filtration is needed to minimize accumulation of scale, bacterial biofilms and other substances on the quartz glass tubes of UV lamps. The major advantages of UV radiation include lack of recognized disinfection by-products, no substantial changes in water quality, no effects on infrastructure and ease of installation. The major drawbacks are that UV light does not provide any residual protection and it does not penetrate well into biofilm, so it should not be used as the sole method for disinfection of a water system (Lin et al. 1992). The action of UV light as a bactericidal method is limited to a distance of 3 cm from the source and the wavelengths emitted by the lamps change with time: when this is combined with the need to install disinfection units near the point of use and maintain them, this method may be costly and high maintenance to be effective. The relatively common practice of adding a UV barrier at the treatment plant prior to entry of the water into the distribution system has not been evaluated for effects on *Legionella* colonization in the distribution system.

Silver-Copper Ionization

Silver-copper ionization is increasingly being tested as a method to control *Legionella* especially at distal outlets in hospital water systems (Rohr, Senger, and Selenka, 1996; Liu et al. 1998; Rohr et al. 1999). The method is based on the electrochemical generation of Ag^{+} and Cu^{2+} cations, which can be bactericidal, and offers interesting features: (a) accumulation of copper in the biofilms and; (b) long-term inhibition of *Legionella* growth (remnant effect). Finally, silver-copper ionization has an action on algae and protozoa as well. Although studies demonstrate conflicting results with regards to copper efficiency, the silver ions are sometimes observed to bring about a concentration-dependent decrease in the level of *Legionella* contamination. Suggested concentrations necessary to inhibit *Legionella* growth are 0.02 to 0.08 mg/L (silver) and 0.2 to 0.8 mg/L (copper). However, this method is often seen as difficult to control because concentrations of copper and silver are usually highly variable within the same system. Moreover, it would have to be proven effective in long-term control of legionellae.

Ozonation

Ozonation is a high-cost method with good local impact but no residual effect on *Legionella* because ozone decays rapidly. Necessary concentrations to get lethality of *Legionella* are 1-2 mg/L. This method is normally combined with several other barriers used during conventional water treatment to be effective, but colonization of drinking water distribution systems with protozoan hosts of legionellae may to some extent negate its effect.

Chlorination and Hyperchlorination

Legionellae have relatively high chlorine resistance (Pankhurst et al. 1990). Both agar-passaged and non-agar-passaged strains of *L. pneumophila* have been shown to be more resistant to chlorine than the coliforms typically used to indicate sanitary quality of potable water (States et al. 1993). This is true even in the absence of host protozoa. The mechanism for chlorine resistance is not known, but might be due to changes in the cellular envelope or may be provided from the other microbiota in the natural environment (Kuchta et al. 1984; Kramer and Ford, 1994). Chlorination remains the most used disinfecting method. Temporary control of *Legionella* colonization is usually obtained with residual free chlorine concentrations of 2 to 4 mg/L. It is estimated that the chlorine concentration needed to decrease their number is 3 to 6 mg/L free chlorine. However, it is not feasible to retain these concentrations in drinking water for other reasons, and legionellae are commonly found in chlorinated distribution systems.

In spite of the natural resistance of legionellae to chlorine, it has been demonstrated that delivery of elevated concentrations of chlorine (e.g. hyperchlorination – where free chlorine concentration is held at 15 mg/L for 24h or 50 mg/L for 12h) in combination with temperature elevation may result in partial and temporary decontamination of hot water systems (Bornstein et al. 1986). Hyperchlorination is therefore practiced quite commonly in hospitals to control the levels of legionellae in hot water systems. However, even during hyperchlorination, legionellae within amoebae may be protected from high levels of chlorine, thus making eradication difficult (Kuchta et al. 1993). It has also been shown that *Legionella* may be engulfed - and then protected - into amoebic cysts that are known to be naturally resistant to high levels of chlorine.

Chloramination vs. Chlorination

Chloramination has been used for drinking water disinfection since 1916. It is often considered as an alternative to free chlorine for water disinfection although a typical monochloramine-using water treatment plant uses free chlorine for initial disinfection and then adds ammonia for formation of the monochloramine for residual disinfection (Kool, Carpenter, and Fields, 1999).

The disinfecting action of monochloramine is slower than that of free chlorine but gives a more stable disinfectant residual throughout the distribution system (Kool, Carpenter, and Fields,

1999). In addition, it has been shown to penetrate better into biofilm than free chlorine. In a recent investigation it was shown that water samples from three municipalities that used free chlorine for disinfection tested positive for the presence of *Legionella* whereas four municipalities using monochloramine disinfection were negative, thus encouraging water utilities using chlorine disinfection to switch to chloramination (Kool, Carpenter, and Fields, 1999). Even if such a correlation was found to be true, the mechanism for chloramine disinfection of legionellae is not known. One could perhaps speculate that the principal action of chloramine may be on the protozoan hosts for legionellae in order to prevent system colonization. However, even chloramine will decline at distal ends of distribution systems and sampling at the consumers tap can show the presence of legionellae in tap water. Moreover, there would be a need to evaluate whether 'removal' of legionellae would really be a net benefit if other microbiological problems are created by the use of chloramine when compared to chlorine. Clearly more studies of the distribution system are needed in this regard.

Thermal Disinfection (Superheating and Flush)

Maintenance of high operating temperatures within a hot water system is generally successful in the elimination of *L. pneumophila*. Some bacteria, possibly in protozoan cysts, may remain in the system despite the elevated temperature (Struelens et al. 1992). In France, it is considered that water temperatures of 70°C are needed during 30 min (flush) at each faucet to ensure a proper, albeit temporary, disinfection of the system. Within the U.S., it is estimated that water needs to be heated to 70°C at the central facility and, during the flush, a temperature of 60°C should be maintained at each faucet. Within hospitals, both in Europe and in North America, this procedure is often combined with chlorination.

In a hospital associated with sporadic infection with *L. pneumophila*, the hot water was maintained at 5 to 6°C above the thermal threshold for suppression of *Legionella* multiplication. Superheating was also performed but both measures failed to eradicate the microorganism (Visca et al. 1999). An elevation in water temperature is likely to select more thermo-tolerant bacterial populations in a microbial consortium. *Legionella* are known to be moderately thermophilic and are likely to be part of the remaining microflora after a thermal treatment if it is not conducted at a high enough temperature. This thermal selection confers a natural advantage for *Legionella* to colonize water systems - *Legionella* colonies were in their highest numbers at 40°C making up about 50% of the total biofilm at this temperature (Rogers et al. 1994b). This is important in systems where the water may not be maintained at a high temperature and therefore increases the potential for infection.

Current Recommendations for *Legionella* Control

In most cases, the reduction in numbers of legionellae to acceptable levels in a contaminated system is only possible with a combination of different measures. Among them are changes in construction and maintenance, as well as, thermal or chemical disinfection, and UV irradiation. The most common method to discourage the colonization of water distribution systems by legionellae is to combine hyperchlorination with an increase in water temperature to above 50°C. Variations in chlorination levels, contact times, water temperatures, and cleaning schedules have been explored and tested. Recommendations include adding 20 to 50 mg chlorine/L to water tanks, maintaining this concentration for 2 hours and allowing circulation until tap water reaches 2 mg of free chlorine/L. It is also recommended that water be heated to 70°C in the tank to obtained 60°C at the water taps. These procedures should be performed at least once a year, before starting a new system or after shutdowns and after an actual or suspected outbreak of legionellosis.

Cultural Characteristics and Nutritional Requirements

The genus *Legionella* consists of thin, faintly staining, Gram-negative, non-spore-forming bacteria (Brenner, Feeley, and Weaver, 1984; Harrison and Taylor, 1988b). They can be grown on L-cysteine- and iron-supplemented BCYEα media (Harrison and Taylor, 1988b). Some species grow well in humidified air at 35°C and others grow better in 2.5 to 3% CO_2 (Harrison and Taylor, 1988b). The average incubation time required for their initial isolation is 3 days, with a range of 1 to 10 days (Harrison and Taylor, 1988b; Winn, 1988; Winn, 1995).

Detection Methods and Criteria for Their Selection

Currently, analyses of environmental samples for *Legionella* involve concentrating the microbial populations and directly plating the concentrates on appropriate recovery media (Feeley et al. 1978; Feeley et al. 1979; Dennis, Taylor, and Barrow, 1981; Edelstein, Snitzer, and Bride 1982; Fitzgeorge and Dennis, 1983; Lee et al. 1993). Although its real sensitivity is unknown, cultivation remains to date the standard for detecting *Legionella* in environmental or clinical specimens. A combination of several media ensures the detection of many known *Legionella* species, except LLAP's, now classified as *L. fallonii*, *L. dronzanskii* and *L. rowbothamii* (Adeleke et al. 2001); these require special culture conditions for extracellular growth. In concentrating legionellae from multi-liter water samples, other more numerous and faster-growing bacteria and fungi are co-concentrated. This makes it difficult to detect legionellae and requires the use of additional sample treatment steps to inactivate or suppress the growth of the contaminating organisms.

Direct fluorescent antibody (DFA) staining can be used for the detection of legionellae in concentrates of clinical samples, but this is often not feasible in dealing with environmental samples because of cross-reactions with members of the genera *Pseudomonas, Xanthomonas* and *Flavobacterium* (Harrison and Taylor, 1988a; Yamamoto, Hashimoto, and Ezaki, 1993). Further, the procedure lacks sensitivity (Yamamoto, Hashimoto, and Ezaki, 1993) and cannot differentiate between live and dead cells (Harrison and Taylor, 1988a). It is also time-consuming and not suitable for screening large numbers of samples.

PCR is often used for the surveillance of contamination of anthropogenic water systems by Legionellaceae or for clinical diagnosis (Starnbach, Falkow, and Tomkins, 1989; Bej, Mahbubani, and Atlas, 1991; Jaulhac et al. 1993; Loutit and Tompkins, 1993; Martin, Fields, and Hutwagoner, 1993; Miller et al. 1993; Nowicki et al. 1993; Paszko-Kolva et al. 1993b; Catalan et al. 1994; Maiwald et al. 1995; Catalan et al. 1997). The problems in the use of certain commercial PCR kits have been mentioned previously. Other PCR assays are available to detect *Legionella* in environmental samples. Multiple primers have been described in order to successfully detect *L. pneumophila* and other *Legionella* species (Jaulhac et al. 1993; Yamamoto, Hashimoto, and Ezaki, 1993; Catalan et al. 1994; Catalan et al. 1997; Miyamoto et al. 1997), especially in water. The use of nested PCR assays is known to increase the sensitivity of the methods to a level nearly identical to that of conventional-PCR assays followed by hybridization on membranes (Miyamoto et al. 1997). This is not the only advantage to these methods: recent results suggest that the effect of potential PCR inhibitors in samples might be reduced because of the dilution of the sample in the second-step PCR (Miyamoto et al. 1997).

Studies are currently underway to assess the potential of newer localization techniques such as fluorescent *in situ* hybridization (FISH) for the detection of legionellae.

Another promising method for the isolation of *Legionella* is immuno-magnetic separation (IMS). This method uses super-paramagnetic polystyrene beads (2.8 to 4.5 μm in diameter) pre-conjugated to a group-specific monoclonal antibody. The resulting antibody-bead complex can recover or selectively enrich microorganisms from environmental samples (Whitmore and Sidorowicz, 1995). However, three main requirements are necessary to develop the immuno-capture technique: (i) to obtain an antibody to a suitable cell surface component, (ii) to prevent non-specific attachment of cells to the beads, and (iii) to obtain bead-cell complexes that are sufficiently stable to be attracted by a magnet (Morgan et al. 1991). The macrophage infectivity potentiator protein (Mip), which is expressed at the surface of the bacteria, can be used as target site that responds to the first requirement because of its role in the initiation of the intracellular infection of both protozoa and human macrophages.

Difficulties known to hamper bacterial detection using IMS include: (a) prediction of the size of the target population in the sample and thus, providing the optimal ratio of beads to sample in advance (Morgan et al. 1991); and (b), the background contamination may be such that an increased number of washes is often needed. This requires a strong antigen-antibody complex able to tolerate the shear forces present during washing steps (Morgan et al. 1991).

Identification Methods and Criteria for Their Selection

Since the first description of the family Legionellaceae in 1979 (Brenner, Steigerwalt, and McDade, 1979), 48 species have been described (Lo Presti et al. 2001). Conventional phenotypic methods for bacterial identification are not suitable for *Legionella*. DFA staining is time-consuming and not suitable for testing large numbers of samples. Other biochemical methods, such as measurements of ubiquinone and fatty acid contents (Moss et al. 1983; Gilbart and Collins, 1985; Marmet, Bornstein, and Fleurette, 1988; Lambert and Moss, 1989; Mitchell and Fallon, 1990), are less frequently used because materials needed to complete the analysis are not readily available. The biochemical methods listed above do not allow the differentiation of all species. Moreover, these methods are often time-consuming and are hampered by the need for specialized pieces of equipment.

Following strain isolation, species identification, in specialized laboratories, is currently based on molecular methods (Fry et al. 1991a; Fry et al. 1991b; Lo Presti et al. 1998b; Ratcliff et al. 1998; Riffard et al. 1998). Indeed, 16S rDNA sequencing (Hookey et al. 1996), *mip* sequencing (Ratcliff et al. 1997; Ratcliff et al. 1998) or DNA/DNA hybridization (Brenner, 1986) are sometimes the sole methods allowing the recognition of the species. Other genotypic methods such as ribotyping (Grimont et al. 1989), arbitrarily primed PCR (AP-PCR) and random amplified polymorphic DNA analysis (RAPD) (Lo Presti et al. 1998b) have been successfully proposed for differentiation of *Legionella* at the species level. Ribotyping is nonetheless a time consuming method not suitable for routine identification of *Legionella* (Grimont et al. 1989), while AP-PCR and RAPD are hampered by their lack of reproducibility (Meunier and Grimont, 1993). Amplification polymorphism of the inter-genic 16S-23S spacer regions (ISR) has been used for identification at the species level in various genera (Jensen, Webster, and Straus, 1993; Gürtler and Barrie, 1995; Harrell, Anderson, and Wilson, 1995; Normand et al. 1996). Consensus primers designed from multiple alignment of 16S and 23S genes of diverse bacteria (Normand et al. 1992; Normand et al. 1996) were successfully tested for their ability to differentiate all *Legionella* species including the bluish-white auto-fluorescent species (Riffard et al. 1998).

Study Objectives

The overall objective of this AwwaRF-funded study is to provide better data on the occurrence of *Legionella* in groundwaters. This information will then be useful for considering whether these opportunistic pathogens should be added to the list of groundwater contaminants. Also, this study is expected to demonstrate that absence of fecal indicators will not correlate with absence of legionellae in groundwaters. Additionally, other physical and chemical parameters were selected to investigate as potential indicators of *Legionella* presence in groundwater. These parameters included, but were not limited to, temperature, pH, dissolved oxygen, turbidity, and iron, and manganese concentrations.

The project was designed as a two-phase study. The purpose of Phase I (pilot phase) was to demonstrate that *Legionella* could be isolated from groundwater samples using the experimental approach proposed by the research team. If successful, the study would enter Phase II to conduct a broader survey on the presence of *Legionella* in samples of water and biofilms from groundwater sites from several locations in U.S. and Canada. This report describes the results from both phases of the study. Despite the common perception that only warm waters are conducive to the growth of *Legionella*, Phase II of this investigation collected and analyzed groundwater samples from both cold and warm regions.

The specific objectives of the Phase I: Pilot Investigation were to:

1. Determine warm-water sample sites to conduct the pilot investigation.
2. Review groundwater temperatures in the United States and Canada and select a range of locations for soliciting groundwater samples for Phase II.
3. Develop and implement a rigorous QA/QC plan.
4. Develop and implement a pilot experimental plan to demonstrate that the presence of *Legionella* can be determined by the sample collection and analytical methods selected for the investigations.

The specific objectives of the Phase II: Assessment Study were to:

1. Collect and analyze groundwater and biofilm samples from a range of sites for the presence of *Legionella* and for a limited number of microbial indicators of fecal contamination.
2. Compare results of groundwater sample analyses for *Legionella* and the selected microbial indicators.
3. Analyze the *Legionella* sample results with respect to groundwater temperatures, well depth and other potentially influential parameters.
4. Design an immuno-magnetic separation assay (IMS) (Olsvik et al. 1994; Whitmore and Sidorowicz, 1995) in an attempt to selectively separate *Legionella* cells from other microorganisms or fragments in the samples to facilitate *Legionella* detection.
5. To determine how widespread legionellae are in groundwater and to identify any factors that could possibly indicate their presence.

CHAPTER 2
MATERIALS AND METHODS

BACTERIAL REFERENCE STRAINS

L. pneumophila (Bellingham-1, ATCC 43111), *L. micdadei* (Tatlock, ATCC 33218), *L. rubrilucens* (WA-270-C2, ATCC 35304) and *L. steigerwaltii* (strain SC-18-C9, ATCC 35302) were used as reference strains for conventional processing and developing immuno-magnetic separation protocols. Strains were conserved in sterile sheep blood as suspensions and were stored frozen at -80°C and used whenever necessary. The Central Public Health Laboratory (Toronto, Ontario, Canada) provided the *Legionella* strains used in this study. *Pseudomonas aeruginosa* (ATCC 49266) was used as specificity control for the Mab 22/1 primary monoclonal (IMS testing).

SAMPLE COLLECTION

Samples for both study phases were collected by the participating utilities in the United States and Canada at or as close to the wellhead as possible. Water samples (3L or 4L depending on the utility) were collected in sterile 1L polypropylene bottles. Biofilm samples were either scraped at the source by the participating utility, or shipped intact and scraped at the laboratory in Ottawa. Biofilm samples came from PVC pipe sections, water meters, coupons, or end-cap scrapings depending on availability at the participating water utilities. Independent of the biofilm type, all biofilm samples were shipped in water to keep the biofilms from drying out in transit. Water and biofilm samples were shipped to the laboratory, on ice, by overnight priority so that they could be processed within 24 hours following sampling.

When using PVC pipe, the sections were attached to a tap and water was allowed to run through them continuously for approximately 7, 15, 30, or 60 days (depending on location and water temperature). The length and diameter of the pipes varied. They were threaded with male adapters at both ends. A Dole valve was installed in front of the pipe to regulate the flow. Flow was regulated to allow water at 2-gallons/min. A drain hose was attached at the end of the pipe to divert the flow to waste. At utility A, for the first samples in Phase II, biofilm was grown for 7 days at between 25 to 30°C. Pipe sections were then removed, refilled with well water (to prevent the biofilm from drying), sealed with end caps and sent to our laboratory on ice. It was subsequently decided to increase the duration for biofilm development to 15 days and then to 30 days. Whenever possible, two connected pipe sections were connected to the well sample tap. The first one was collected after one month and the second one after two months of colonization. This was done since it is not known if *Legionella* are pioneering bacteria in the development of biofilms. Thus not all the biofilms samples tested were of equivalent age – varying from 1 week to several years.

A questionnaire (Appendix B) was submitted to each participating water utility to collect as much additional information as possible from each utility. Whenever possible, in-house chemical analyses of the samples processed were collected to examine the likely role of these variables on the presence or absence of *Legionella*.

SAMPLE PROCESSING

Water Samples

Upon receipt of water samples, each liter of water (usually 3L) was separately filtered through sterile 47-mm filter funnel assemblies containing 0.45-μm porosity polycarbonate membrane filters. After filtration, filters were immediately removed aseptically and placed in separate 50-mL glass vials containing 10-mL sterile water. When more than one filter was necessary per L, all the filters for that L were added to the same vial and, for each additional filter, a 10-mL volume of sterile water was added. All filtered samples were then mixed with a vortex mixer (3 x 30 s) to re-suspend the bacteria.

It has been reported that incubation of water samples containing amoebae for one week at 35°C improves detection of legionellae by the culture method (Sanden et al. 1992); *L. anisa*, *L. bozemanii*, *L. gormanii*, *L. micdadei*, *L. pneumophila* (serogroups 1, 2, 3, 5, 6, 7 and 10), *L. rubrilucens*, *L. sainthelensi* and *L. steigerwaltii* have been isolated after incubation of samples that were negative by conventional procedures. Thus, without knowing if *Legionella*-containing amoebae were in the samples, we decided to incubate one filtered aliquot equivalent to 1L of water at 35°C for 1 week before attempting cultivation. Due to massive overgrowth in cultures from these 'enriched' fractions, this procedure was modified during Phase II so that it was used only to enrich the *Legionella* for PCR processing by releasing potential intracellular strains. For selected samples in Phase II only, an additional filtered aliquot equivalent to 1L of water was sent for protozoa analysis.

The other filtered aliquots were subsequently sonicated at low power in a sonicating water bath for one minute to further break up clumps and release any intra-amoebal legionellae (Boulanger and Edelstein, 1995). Glass tubes are favored over plastic for better transfer of ultrasonic energy. One of the sonicated and vortexed samples equivalent to 1L of water was used for cultivation on the day of its arrival, the other was used for PCR analysis.

Biofilm Samples

Biofilms not scraped and resuspended by the utilities, were scraped and resuspended in the laboratory in the water in which they had been transported. If this volume was too small, sterile deionized and distilled water was used to make up to a minimum resuspension volume of 20 mL. Any clumps of organisms and/or their aggregates as biofilms were dispersed by vortexing the suspensions at full speed (3 x 30 s) (Rogers et al. 1994a; Rogers et al. 1994b; Kalmbach, Manz, and Szewzyk, 1997). One third part of the sample was stored at 35°C for 1 week to enrich the *Legionella* for cultivation (Phase I) and PCR processing (Phase I and II). An additional part was sent for protozoa analysis (Phase II). The rest of the volume was dispersed by mixing with a vortex mixer (3 x 30 s) and by the use of low power sonication in a sonicating water bath for 1 min to release intra-amoebal legionellae (Boulanger and Edelstein, 1995). The sonicated and vortexed fraction was used for cultivation on the day of its arrival and for PCR analysis.

CULTIVATION AND IDENTIFICATION OF LEGIONELLAE

In view of the known difficulties of recovering low numbers of legionellae from environmental waters, a relatively complex cultivation scheme was devised to allow each sample to be subjected to three different types of treatment prior to being plated on three types of media (with and without antibiotics) at two incubation temperatures and in duplicate. As discussed earlier, the use of preliminary treatment(s) and a selective medium was necessary to partially inactivate the growth of those organisms in the specimens that were co-concentrated and which could interfere with the recovery of *Legionella* spp because of overgrowth or inhibition (Toze et al. 1990).

The treatments used were – untreated, heat treated (Edelstein, Snitzer, and Bridge, 1982) and acid treated (Bopp et al. 1981; Bueschung, Brust, and Ayers, 1983); *Legionella* is able to survive at elevated temperatures and is resistant to several harsh conditions, including low pH, and can be selected for through these means. For heat treatment, 2 mL of the sample was placed in a 50°C water bath for 30 minutes. The samples were allowed to return to room temperature before cultivation on semi-selective media. For acid treatment, 1 mL of the sample and 1 mL of the acid treatment reagent were combined (giving a pH of 2.2) and allowed to stand for 15 minutes. These samples were then neutralized to pH 6.9 before cultivation on semi-selective media. It should also be noted that either or both of these treatments may cause a loss in viability of legionellae. The degree to which this occurs could be a function of the prior conditions to which any legionellae have been exposed and their particular sensitivities. One study (Boulanger and Edelstein, 1995) reported a loss in bacterial viability for acid treatment of approximately 30%.

The cultivation of *Legionella* was performed using three different solid-phase media, all based on Buffered Charcoal Yeast Extract (referred to as BCYEα) (Feeley et al. 1978; Feeley et al. 1979; Edelstein, 1982; Edelstein, Snitzer, and Finegold, 1982). GVP is BCYEα that has been supplemented with glycine, vancomycin, and polymyxin B (Wadowsky and Yee, 1981). CCVC is BCYEα supplemented with cephalotin, colistin, vancomycin and cycloheximide. The semi-selective CCVC and GVP media were used to try and further reduce the bacterial load during cultivation. Freshly prepared plates of the media were allowed to equilibrate to room temperature for at least 2 hours before their inoculation. The pH of the prepared media was also checked as explained in the QA/QC control section. Anisomycin was used as a supplement for GVP during Phase I only. Due to its cost, it was decided that anisomycin would be no longer used during Phase II.

One hundred μL of each sample (direct plating - DP -, acid plating - AP -, and heat treatment - HT) was inoculated onto each of the three different media in duplicate and incubated at 35°C or 30°C in a humidified (>50%) 2.5 to 5% CO_2 atmosphere for up to 14 days. The duration for initial recovery of these bacteria may require 1 to 2 weeks or more. Incubators were checked daily for temperature and % of CO_2 (QA/QC control). Each inoculated plate was left at room temperature for approximately 1 hour before placing upside down for incubation under the conditions described above. The conventional culture procedure incubates plates for *Legionella* recovery only at 37°C. However, many of the legionellae in these samples may be much more accustomed to lower temperatures and may therefore be better able to grow at lower temperature. Also lower temperatures of incubation are sometimes more suited for injured microorganisms and legionellae in these samples could be injured through prolonged exposure to disinfectants if these are injected into the well or through the pretreatments that the samples were subjected to. The incubation of plates at lower temperature (i.e., 30°C) may also facilitate the recovery of atypical *Legionella* species (Newsome et al. 1998). It has been speculated that the inability of some strains or species to multiply at 37°C under laboratory conditions might be correlated with a less virulent phenotype. However, these strains or species might be more virulent for protozoa at lower temperature resulting in a better spreading of the bacteria in the environment.

Quality control was performed with bacterial strains that have been passaged minimally on agar. It should be noted that laboratory-adapted strains, including some available from the American Type Culture Collection (ATCC), grow well on many types of media that may not support the growth of field strains.

Each plate was viewed through a dissecting microscope every 72 hours. *Legionella* colonies have a "mottled" surface, an iridescent red-blue-green shed, or a faceted "cut-glass" appearance. Although the colonies themselves are round and entire, the play of light on the surface produces an appearance of irregularity or complex internal structure. As colonial growth continues, the iridescence can be more apparent with the naked eye, particularly where growth is confluent. In short, the observation of opaque bacterial colonies that have a "ground-glass" appearance should be considered as suspected legionellae. Plates with *Legionella*-like colonies were placed in a biological safety cabinet. Each suspect colony was aseptically picked and inoculated onto BCYEα agar, a BCYEα agar plate without L-cysteine and a blood agar plate and incubated for 48 h. Plates without growth on either BCYEα without L-cysteine or blood agar were re-incubated for an additional 24 h. All colonies were checked for auto-fluorescence under long-wavelength UV light. Isolates demonstrating growth on only BCYEα agar were presumptive for *Legionella*. Care should be taken when the L-cysteine requirement for growth is used as a differential criterion for the presence or absence of *Legionella* in a sample. Some strains are able to grow on BCYEα without L-cysteine even on their first isolation from environmental samples (Orrison et al. 1983). Moreover, it seems that some species (for example *L. bozemanii, L. dumoffii, L. gormanii, L. jordanis, L. longbeachae, L. micdadei, L. oakridgensis, L. pneumophila*) no longer require iron after several passages on BCYEα (Orrison et al. 1983). Finally, some *Legionella* species are able to grow on sheep blood agar exhibiting the typical "ground-glass" appearance (Lo Presti et al. 1998a). Should there be a presumptive colony; DNA isolation followed by PCR was carried out. Partial nucleotide sequencing of the 16S rRNA gene was used for identification of the *Legionella* species.

Calculations/Expression of Results

It was desirable to try and quantitate *Legionella* recovery from the samples, even if this could only be an approximation. By convention, one colony forming unit (CFU) is considered to come from one viable bacterial cell. However, with 3 treatments, 3 media and two incubation temperatures, all of which in Phase 1 were performed on both enriched and non-enriched samples, there would have been 36 separate numbers. Not only would each of these numbers have been obtained from a maximum of 0.2 mL, representing at best 20 mL of original sample, but their accuracy would have been marred by the considerable degree of overgrowth experienced on many of the plates. A total volume of 3.6 mL was inoculated, corresponding to approximately 1/3 of the sample or 350 mL when the eluate volume was 10 mL. For larger volumes of sample concentrates it corresponded to less of the original sample.

When legionellae were detected it was quite common to find them on several plates but this did not always occur. One way of simplifying the presentation of quantitative results would be to take a mean value of the counts on all plates, and this was done for Phase I.

The concentration of legionellae in each sample was therefore calculated using the following equation

$$\mathbf{N} = \mathbf{n} \times 10 \times \mathbf{C} / \mathbf{V}$$

Where:

N is the number of *Legionella* per L of sample (CFU/L)

n is the number of colonies identified as *Legionella* on a single plate, or **n** = n1 + n2 + n3 / 3 where *Legionella* colonies are found on multiple plates and the mean is employed for calculation

n1 = Sum of the number of colonies on DP plates / number of plates

n2 = Sum of the number of colonies on AP plates / number of plates x D

n3 = Sum of the number of colonies on HT plates / number of plates

C is the volume of concentrated sample in mL (either water or biofilm sample). For water samples, **C** = 10 mL if one filter has been used to concentrate the sample (that was most of the time).

V is the volume of filtered sample (in the case of water)

D is an additional dilution factor for acid treatment - usually **D** = 2.8 (1 mL of sample after filtration + 1 mL acid buffer + 0.8 mL neutralizing buffer).

However, this was considered to significantly underestimate the legionellae present and artificially give each treatment equal weighting which may not be a valid assumption. Therefore, for Phase II, a different approach was taken to getting a single number of legionellae in the sample. The following should be noted:

- Each treatment could have distinctly different effects on the legionellae present.
- Each media will allow a different mixture of legionellae and competing organisms to grow.
- Even within the pair of duplicate plates, large differences could sometimes occur, especially if one of the plates was more overgrown than the other. For example, it may have been possible to enumerate and isolate legionellae from the less overgrown plate but not from the more overgrown one. This does not mean that legionellae were necessarily absent in the other plate.
- Overall, there were a large number of completely or partially overgrown plates which could have obscured *Legionella* colonies, or prevented them from growing due to antagonistic organisms. Detection of the *Legionella* colonies was quite tedious and difficult and some could be easily missed.

It was generally considered that any legionellae detected were an absolute minimum of those present. It was therefore decided in Phase II to treat each plate individually and calculate the numbers of legionellae in the samples from the individual plate showing the highest number of *Legionella* colonies regardless of the type of treatment. Whichever of these two methods is used, it needs to be borne in mind that the numbers of legionellae quoted are estimates only.

In the case of biofilm samples, **C** corresponds to the total amount of biofilm + sediments resuspended in a known volume of raw water and obtained from pipe sections or scraped from the wall of the wells depending on sampling methodology chosen. Here also it was considered desirable to express results differently (i.e., whenever possible by the number of CFU/cm^2) during Phase II of the study in order to get more meaningful and comparable results from different sources of biofilms.

Variations in Sample Processing Between Phase I and Phase II

The sample processing procedure varied slightly between Phase I and Phase II, and this is shown in Table 2.1. The experimental approach to cultivation and PCR is summarized in Figure 2.1 with specifics noted for Phase I. A general outline of the overall sample processing, including cultivation, PCR and IMS, is shown in Figure 2.2. Specific notes on cultivation in Phase II are shown in Figure 2-2.

Table 2.1
Comparison of cultivation procedure for Phase I and Phase II

Procedure	Phase I	Phase II
Biofilm collection	All biofilm samples were collected by removing end-caps, scraping biofilm from cap, and resuspending in water from the well	Biofilm samples were collected on PVC pipe sections, water meters, coupons, or as was done during Phase I
Water collection	3 to 4 liters of water were collected in 1 liter bottles	3 to 4 liters of water were collected in 1 liter bottles
Shipment to laboratory	Overnight delivery in a cooler on ice	Overnight delivery in a cooler on ice
Sample concentration	Each 1 L of water was concentrated to 10 mL Biofilm samples were plated directly	Each 1 L of water was concentrated to 10 mL Biofilm was scraped from pipes, meters or coupons, suspended and accompanying water and plated (End-caps were done as in Phase I)
Sample plating	Each sample was plated (on the day received) onto 3 media using 3 separate treatments (total of 30 plates per sample) In addition, each sample was incubated at 35°C for 1 week and plating was done as stated above.	Each sample was plated (on the day received) onto 3 media in quadruplicate using 3 separate treatments (total of 36 plates per sample) The one-week incubation for sample enrichment was not done
Legionella detection and identification	Each plate was viewed under dissecting microscope every 3 days for 2 weeks, suspect colonies were subcultured for presumptive identification, PCR and DNA sequencing was used for positive identification	Each plate was viewed under dissecting microscope every 3 days for 2 weeks, suspect colonies were subcultured for presumptive identification, PCR and DNA sequencing was used for positive identification
Legionella quantitation	Quantitation of *Legionella* used the equation N = n x 10 x C / V from mean values. Quantitation was also repeated using highest number on individual plates to be compatible with Phase II	Quantitation of *Legionella* used the equation N = n x 10 x C / V from highest number on individual plate

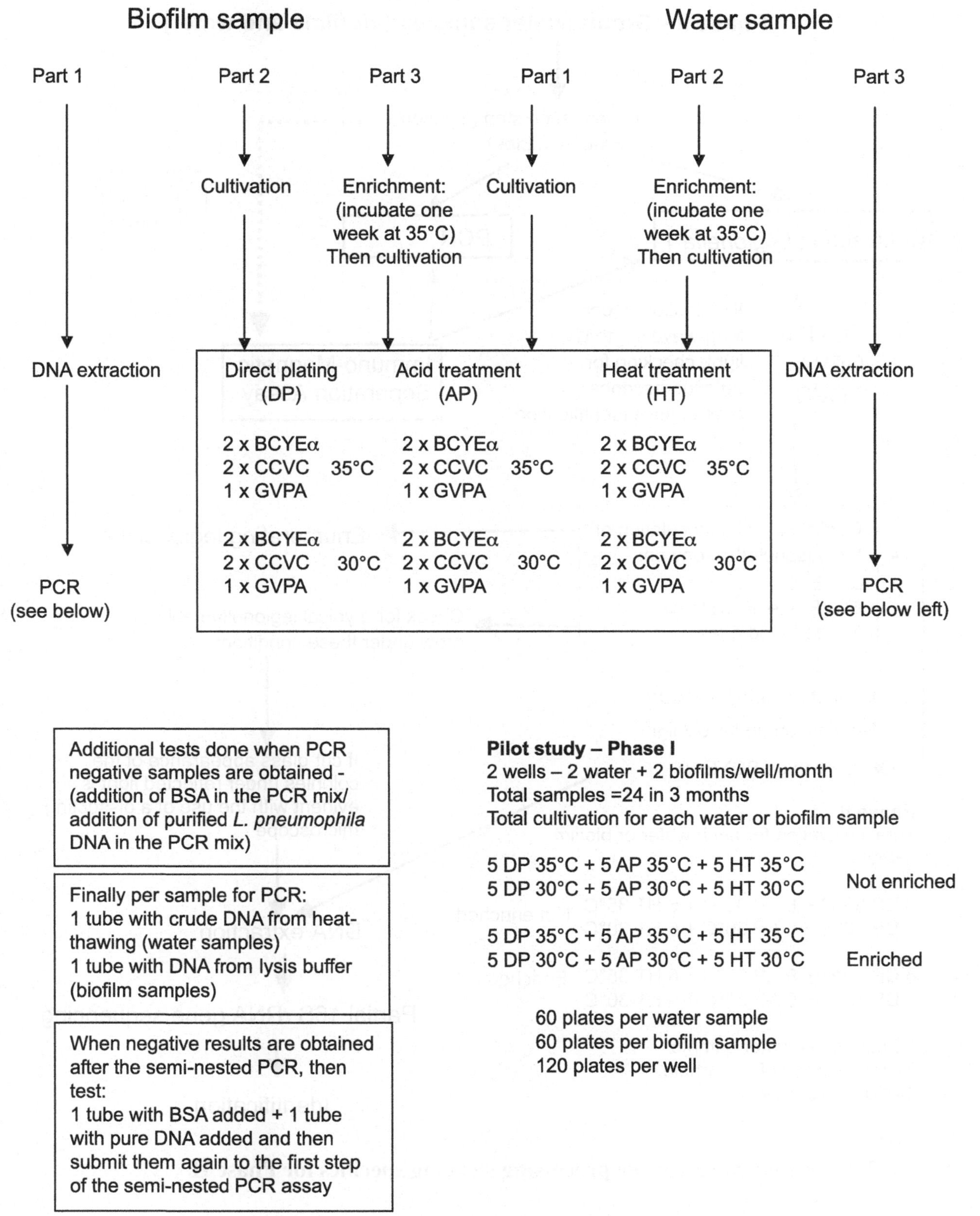

Figure 2.1 Approach to cultivation and PCR for Phase I & II

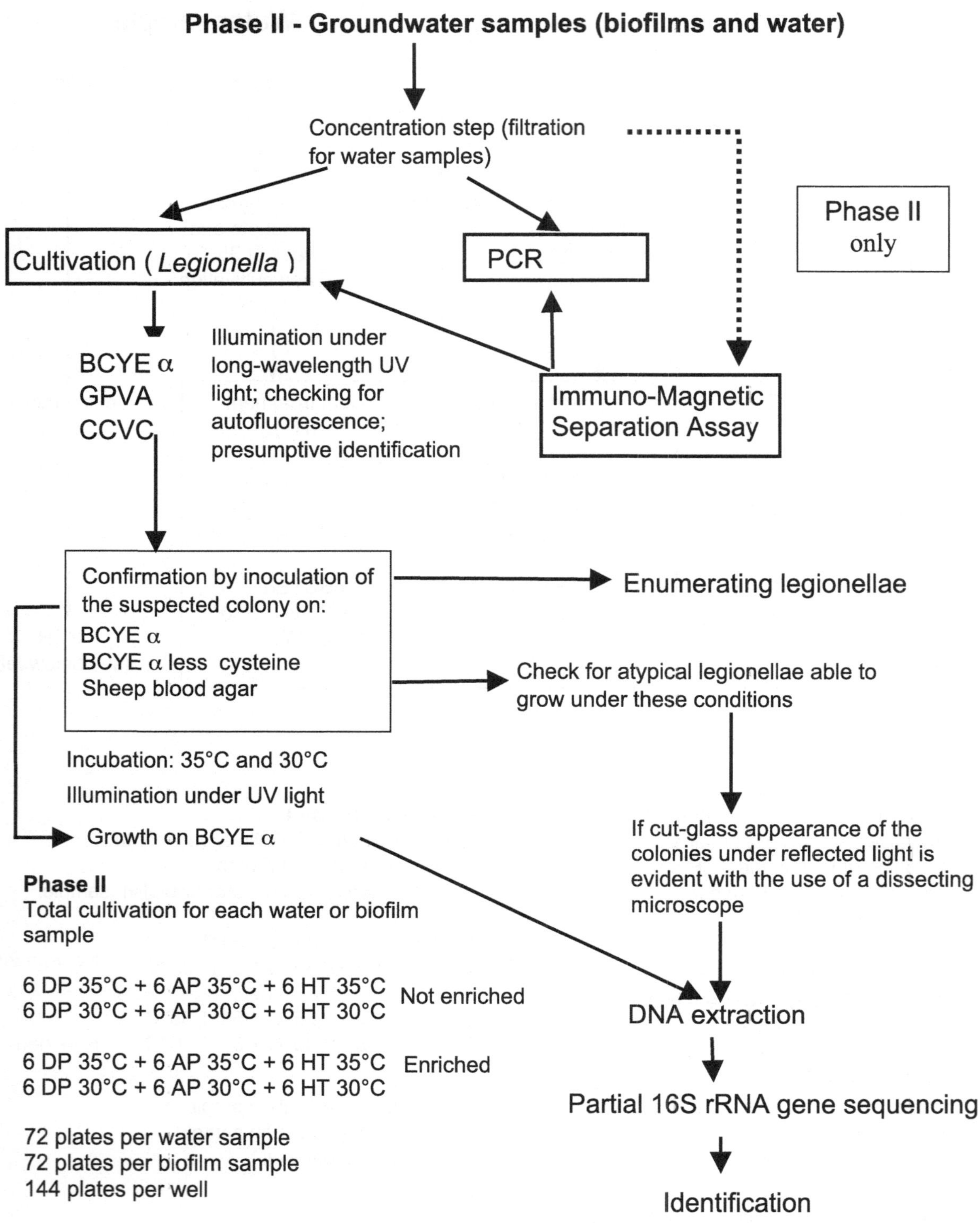

Figure 2.2 Groundwater sample processing showing specifics for Phase II

Table 2.2
Conditions for any additional bacteria testing required

Organism	Media	Incubation temperature	Incubation time	Sample dilutions
Fecal coliform	mT7	44.5°C	24 hours	Undiluted, 1/10, 1/100
Total coliforms	mT7	35°C	24 hours	Undiluted, 1/10, 1/100
Fecal streptococcus	m Enterococcus	35°C	48 hours	Undiluted, 1/10, 1/100
Heterotrophic bacteria count	R2A	Room temperature (22°C)	7 days	Undiluted, 1/10, 1/100

Additional Bacteria Testing on Water and Biofilm Samples

Water quality data requested and used were primarily generated by participating water utilities in their regular monitoring programs. When insufficient data was provided, microbiological parameters were measured in our laboratory as shown in Table 2.2, using membrane filtered samples of 100 mL for water and suitable dilutions of material suspended from scraped biofilms.

Semi-nested PCR Assay

A semi-nested PCR assay, recently described by Miyamoto et al. (1997), was used for this study. This method is useful to detect *Legionella* species from environmental samples (Miyamoto et al. 1997). The method is particularly simple, rapid and has the great advantage of reducing the effect of potential PCR inhibitors (Miyamoto et al. 1997). Its specificity has been successfully tested against 39 *Legionella* species. Its sensitivity allows the detection of one fg of *Legionella* DNA (Miyamoto et al. 1997).

Sample preparation for PCR involved the use of a published DNA extraction method during Phase I (Jaulhac et al. 1992; Jaulhac et al. 1993), and for Phase II, a commercially available DNA extraction kit (Dneasy Extraction kit, Qiagen) was used to reduce the effect of PCR inhibitors. The semi-nested PCR assay, and detection of the amplified products, was performed as described by Miyamoto et al. (1997) and is described more fully in Appendix A.

For samples that were found to be negative by the semi-nested procedure, these samples were then tested for the presence of PCR inhibitors. DNA was added (2 μL or 200 ng of DNA extracted from *L. pneumophila*) to 3 μL of the sample. In this case only the first step of the semi-nested PCR was performed. If no *L. pneumophila* PCR product was seen, the samples were presumed to contain one or more PCR inhibitors. No specific attempts were made in this study to remove PCR inhibitors. This was partly due to sample limitations and partly due to the potential loss of low concentrations of legionellae.

In addition to direct detection of legionellae by PCR, amplification of *Legionella* specific sequences were also used for identification of legionellae as mentioned above, and as discussed in more detail in Appendix A.

Immunomagnetic Separation (IMS) Assays

An immunomagnetic separation (IMS) assay was developed during the course of this study as an alternative, and potentially easier, method for detecting *Legionella* in environmental samples. A genus-specific monoclonal antibody (Mab), specific to the genus *Legionella*, was obtained from Dr Juergen Helbig (Institut für Medizinische Mikrobiologie und Hygiene, Universitätsklinikum der TU Dresden, Dresden, Germany). This antibody (mAb 22/1) is able to recognize a *Legionella*-specific epitope (Helbig et al. 1995). The developed IMS assay was used on selected samples in Phase II to evaluate its ability to enrich for legionellae and enhance their detection from groundwater samples by addition to the more conventional assay methods.

The mouse monoclonal antibody (Mab known as 22/1, IgG1 subclass) able to recognize a *Legionella*-specific epitope of the Mip (macrophage infectivity potentiator) protein (Helbig et al. 1995) was received as cell culture supernatant and was purified using a protein-G column following manufacturer's instructions (Amersham Pharmacia Biotech, Inc., NJ, USA). Sheep anti- mouse IgG1, coated onto Dynabeads® M-450 (Dynal Inc, NY, USA), was used as the secondary antibody to detect *Legionella*-Mab 22/1 complexes following manufacturer's instructions. Non-specific binding assays employed a commercially available affinity purified mouse IgG Fc antibody (Chemicon).

Although Mab22/1 was known to specifically detect *Legionella* (genus-specific), its specificity for *Legionella* detection by IMS was tested by flow cytometry against a strain of *P. aeruginosa* as representative of strains frequently encountered in environmental samples and known to cross-react with *Legionella* when using direct fluorescent antibody typing. One fraction of the purified Mab 22/1 was FITC-labeled using a FluoReporter® FITC Protein Labeling Kit following manufacturer's instructions (Molecular Probes Inc.). An Epics XL™ (Coulter Corp.) cytometer equipped with a 15-mW, air-cooled, 488-nm-wavelength argon laser light allowing a 4-colour fluorescence analysis was used to measure forward angle light scatter (FSC), and fluorescence of the microbial cells. Set up of the flow cytometric assay was done using pure cultures of *L. steigerwaltii* (strain SC-18-C9, ATCC 35302). Photo-multiplier tube voltage and spectral compensation were initially set using cells single-stained with propidium iodide (PI) and FITC-labeled 22/1 antibody. Although resolution was obtained for the strains investigated, flow cytometry was dropped because of the decision that the procedure should be directed towards maximizing cell recovery. As a consequence, following cell-bead interaction, the beads and cell-bead complexes were attracted towards the magnet and held at the side of the tube during the single washing step undertaken. Any disruption of the attracted beads at the bottom of the tube was avoided. Additional washing steps were not done. It was noted that the magnetic beads, even though they are super-paramagnetic, appear to attach to each other through inter-magnetic force making it difficult to put them back into suspension after more than three washing steps. Moreover, large populations of attached cells were removed from the surface of the beads after several washes (data not shown). During the assay, two procedures were tested (combined IMS and uncombined IMS) to find the more sensitive approach.

Uncombined IMS Procedure

The Mab 22/1 (20 μL of a 1 μg/mL suspension) was added to pure bacterial suspensions or artificially contaminated natural samples. Mixtures were left at room temperature for 30 minutes with agitation. Sheep anti-mouse coated Dynabeads® were then added (15 μL of a 30 μg/μL suspension) and gently agitated at room temperature for 30 min. IMS was then performed (target cells were attracted by applying a magnet at the bottom of test tubes for 2 min). One washing step (with 1-mL sterile water) was performed with the cells being attracted on the magnetic support. The concentrated cells were resuspended in 200 µL or 1 mL (depending on the experiment). Controls were done in parallel to determine the recovery rates when pure bacterial suspensions were used (sensitivity).

Combined IMS Procedure

The Mab (10 µL of a 10 µg/mL solution) was added to 1 mL of the sheep anti-mouse coated Dynabeads and incubated with gentle agitation for 24 hours at 4°C. After the 24-hour incubation, the bead-antibody complexes were drawn to the side of the tube using a magnetic support and washed with 1 mL of PBS. The bead-antibody complexes were resuspended in the PBS wash solution and agitated for an additional 15 minutes at room temperature. This washing procedure was repeated with the final resuspension of the beads in 1 mL of PBS. The combined antibodies were stored at 4°C until used. Varying amounts of the combined antibody were added to the sample and allowed to incubate for 1 hour with gentle agitation. IMS was then performed (target cells were attracted by applying a magnet at the bottom of test tubes for 2 min). One washing step (with 1-mL sterile water) with the cells being attracted on the magnetic support was done. The concentrated cells were resuspended in 1 mL. Controls were treated in parallel to determine the recovery rates when pure bacterial suspensions were used (sensitivity).

Finally, IMS was also tested for its ability to reduce the impact of PCR inhibitors during the PCR assay (testing done on selected samples during Phase II only).

Protozoa Analysis

Protozoa analysis on selected samples was performed by Prof. Marilyn Lee, Ryerson University, Toronto, Canada. The detailed protocol used is provided in Appendix A.

Geological, Hydrogeological and Water Quality Parameters

Microbial and chemical measurements for groundwater vulnerability assessment (Abbaszadegan et al. 1998) suggest that it may be possible to evaluate if any association exists between a range of parameters and *Legionella* occurrence. Such associations, if they exist, might be effective in distinguishing between wells at "low-risk" or "high-risk" for *Legionella* occurrence. Therefore, the questionnaire sent to participating water utilities (Appendix B) included a range of questions on the physical characteristics, locations and hydrogeology of individual wells. However, for legionellae, the association may perhaps be more reasonably expected with the protozoa known to support Legionella growth, and to perhaps enhance their long-term environmental persistence.

Additional Data Submitted

Water utility "A," for which regular sampling was performed during this study, had a variety of other data performed independently on the tested wells. PLFA (determination of the phospholipid ester-linked fatty acids) and DGGE (Denaturing Gradient Gel Electrophoresis) analyses were performed by Microbial Insights, Inc. (Rockford, TN). The PLFA method is based on the extraction of lipid biomarkers from the cell membranes and walls of microorganisms. These lipids are then analyzed by gas chromatography and mass spectrometry for identification of each individual lipid. The profile obtained provides:

- a quantitative measure of the viable or potentially viable biomass;
- a quantitative definition of the patterns of the ester-linked fatty acids in a microbial community (such as a biofilm or water microbial community). Thus, statistical analyses of the patterns of whole microbial communities quantitatively define relatedness of different microbial communities;
- physiological adaptations and the growth phase of the bacteria in the community.

The DGGE method uses PCR with universal 16S rDNA-targeted primers in combination with a denaturing gradient gel electrophoresis to separate all the PCR products that have the same length but different sequences. It can be used to identify bacteria to the species level directly from a complex sample (Muyzer, De Waal, and Uitterlinden, 1993; Muyzer et al. 1995; Muyzer et al. 1998; Muyzer and Smalla 1998; Muyzer 1999; Rooney-Varga et al. 1999). Separated PCR products are sequenced and matched to database libraries to determine bacterial species that are present in the analyzed sample. It is often believed that the intensity of each band correlates to the relative abundance of each bacterial species in the sample analyzed. Water utility "A" has kindly provided all these additional analyses.

Statistical Analysis

Statistical analysis was done by Dr. David Abrial using ADE 4.0 ("Analyse de Données Ecologiques," ecological data analysis): a multivariate analysis and graphical display software (Thioulouse et al. 1997). This software allows the use of multivariate analysis methods, spatial data analysis methods, discriminant analysis within/between groups analyses, many linear regression methods, multiple and PLS (partial least squares) regression and orthogonal regression (principal component regression), projection methods like principal component analysis on instrumental variables, canonical correspondence analysis and many other variants, coinertia analysis and the RLQ method, and several three-way table (k-table) analysis methods. ADE 4.0 can be obtained freely by anonymous FTP to http://biom3.univ-lyon1.fr/. The WWW documentation and downloading page is available at: http://biomserv.univ-lyon1.fr/ADE-4.html, which also provides access to updates and user support.

From the three basic multivariate analysis methods available using ADE 4.0 and which can be applied to one-table data sets, the principal components analysis module for quantitative variables was selected and then data were analyzed using simple variance testing and discriminative factorial analysis.

Technical Difficulties and Technical Notes

During the course of this study a variety of technical difficulties were encountered with regard to the sampling, and certain technical issues were noted with a potential to affect detection of legionellae by the methods used. For the sake of completeness, these have been detailed in Appendix A. In addition, technical details for the methods used, and which are not included above, have been added into Appendix A.

CHAPTER 3
RESULTS: PHASE I, PILOT STUDY (PROJECT NUMBER 2594)

GROUNDWATER SELECTION CRITERIA FOR THE PILOT STUDY

As stated in RFP 2594, the major parameter determining well selection for this study is groundwater temperature. Ecological studies have demonstrated that *L. pneumophila* can multiply in water from 25°C (lower limit) (Yee and Wadowsky, 1982) to 42-50°C (upper limit) (Wadowsky et al. 1985). They can also be associated with biotopes where temperatures can rise to 60°C (Fliermans, 1983). On the other hand, *Legionella* are able to survive at low temperatures since they have been successfully isolated from frozen rivers (Fliermans 1983) as well as contaminated ice and water dispensed from ice machines (Stout, Yu, and Muraca, 1985; Graman, Quilan, and Rank, 1997).

The purpose of the Pilot Study (Phase I) was to demonstrate that *Legionella* could be detected using the methodological approach selected by the research team. It was therefore decided to primarily target those warm groundwater systems that were thought to be potentially favorable for *Legionella.* Based on our knowledge of *Legionella* ecology and our review on groundwater temperature within the U.S., we selected participating utilities from **geographical areas with elevated temperatures**. At these locations, with groundwater temperatures normally in the range of 20-30°C, it was hypothesized that even if free-living legionellae were absent, protozoa may indeed be present as bacterial predators and intracellular production of legionellae in the protozoa may be likely.

It is well known that legionellae are frequently associated with surface water. Groundwaters under the influence of surface water can be expected to be those most likely contaminated with legionellae. The research team decided to focus its investigation on **groundwater systems that are not known to be under the direct influence of surface water.** Although the risk associated with this decision was to not detect *Legionella* in the samples, the discovery that *Legionella* colonize groundwater systems that are not known to be under the direct influence of surface water may have significant implications for human health and the approach to water treatment of these public systems. Furthermore, it could point to the potential for unknown surface influences to impact the microbiology of aquifers and well infrastructure.

These two criteria dictated that the Phase I studies focused on utilities in the southern part of the United States. Accordingly, two utilities in Florida were selected. These utilities were designated "A" and "B." Data showed that temperatures of water from their wells was relatively constant, ranging from 24°C to 27°C even when those wells were situated between 300 and 900 feet deep. On the basis of data provided on 13 wells in the wellfields owned by water utility "A," heterotrophic associated microflora counts increased from June to October whereas total coliform and *E. coli* were rarely detected through the survey, suggesting that these wells were not significantly under the direct influence of surface water. However, in spite of their depth the potential for more remote surface influences could not be refuted, and the mostly high TOC data (Tables 3.2 and 3.3) suggest human impacts within the wellfields.

SAMPLING STRATEGIES

The sampling strategies were developed at the individual utilities according to local practicalities. Water samples were collected as close to the well head as possible from appropriate sampling taps. At water utility “A” a dedicated sampling tap was installed on the end cap of the wells due to the regular samples collected. The water was allowed to flow for 15 minutes before sample collection. Field measurements were recorded (water temperature, pH, dissolved oxygen, turbidity, and conductivity). Participating water utilities were provided with a “sample and data sheet” (see Appendix C) to record field data.

Whenever possible, the idea was to collect long-established biofilms. However, they may not be readily available at the wellhead, and use of biofilms from the distribution system could confound the results. Therefore, the exact means of sampling the biofilms was decided on local practicalities in conjunction with the utility management. At water utility “A,” the well water pumps were turned off and the gate valves for the main transmission line to the neighboring wells were closed. Once the pressure in the wells was released, the end caps were removed and the biofilm samples were taken by using sterile scalpels to scrape off the biofilm. Water utility “A” has many wells, and the selection of which ones to use for the pilot study was based on previous studies on and knowledge of this wellfield. Microbiological data, production capabilities (very high to very low production wells) were considered primary parameters. Wells were also selected based on their accessibility and relative ease of sampling.

The strategy to sample wells for water utility “B” was established by the Manager, Monitoring and Mitigation. The process is different from that for water utility “A” because no well casings could be opened to access biofilms. Briefly, there were two possibilities to get biofilms: (1) biofilms could be scraped off from the seals around the production well pump heads (these locations are open to the atmosphere and can be contaminated by whatever is in the air); or alternatively (2) biofilms could be obtained from sections of the feed lines for this sealed area (the feed lines are either 3/4" PVC or galvanized pipe which have been finally removed, refilled with well water and sealed with caps). The second strategy was chosen and the selection of wells to be sampled was decided by the utility.

It is important to keep in mind that the Phase I pilot study was based on one-time sampling events and may not represent the ongoing levels of contamination of the selected wells.

SAMPLES, PARAMETERS AND DATA FROM PARTICIPATING UTILITIES

Water Utility A

Well information and data on field parameters for Water Utility A are shown in Tables 3.1 and 3.2, respectively. The well information was partially collected from the summary report on the 1994 wellfield microbiological monitoring program and from statistical analysis done in 1997. This latter analysis has shown that coliform and color are related in this wellfield, total organic carbon and color are related and coliform and turbidity are related. The increase in coliform bacteria in samples from this wellfield is statistically significant and shows that these organisms are indicative of the presence of biofilms, which are localized in the wells (report on the Assessment of Coliform Occurrence in the water utility “A” wellfield prepared by Charlotte Smith and Associates, Inc. 1997). This could partially suggest that some surface influences were possible but the depth of the wells may suggest otherwise.

Table 3.1
Well information for Phase I and Phase II for water utility "A"

Well ID	Annual well production (million gallons)	Bacteriological data	Other information
PW_5N	89.77 – 242.68	No MMO-MUG failures, BART shows pseudomonad population	3 sections of end cap scraped
PW_113	14.09-113.09	37 of 55 failed MMO-MUG (*Enterobacter cloacae, Pseudomonas stutzeri, Serratia liquefaciens, Citrobacter* species) 4 samples > 500 CFU/mL HPC	Installed in 1961, water temperature constant 25°C, samples obtained during Hurricane Irene
PW_103	169.94-432.24	1 of 12 failed MMO-MUG (*Enterobacter cloacae*) 1.7 x 10^5 CFU/mL HPC BART shows pseudomonad presence	Well possibly under influence of surface water
PW_136	86.05-480.54	Iron related bacteria and pseudomonads (BART) 4 of 29 MMO-MUG failed (*Enterobacter cloacae*)	None
PW_138	34.20-154.59	23 of 42 failed MMO-MUG (*Enterobacter cloacae*) pseudomonads present (BART)	Installed 1966
PW_5S	7.15-129.43	No MMO-MUG failures Pseudomonad population	Situated on hypothesized fault line

Depending on which well was sampled, biofilm and water characteristics varied. During Phase I, all biofilm samples were collected by scraping the end cap from the production well and placing the scrapings into a bottle containing well water. The amount of biofilm development ranged from barely visible to more than 3 mm in thickness and ranged in color from rust to brown to grey to black. The biofilm on many of the end caps that were scraped could be described as crusty. Bulk water samples (4 L per sample), in many cases, appeared clear but in some cases, the water had a slight yellowish color (possibly from iron or organic matter). Also, a rotten egg odor could be detected in some water samples (presence of H_2S).

Since 1988, water utility "A" has tested wells in this wellfield for total aerobic bacteria, iron-related bacteria, sulfate-reducing bacteria, slime-forming bacteria and fluorescing pseudomonads. Each well was also analyzed for the presence of total coliforms/*E. coli* every month using MMO-MUG (P/A, 35°C/24h, Colilert) using 250 mL of raw water. If the presence of total coliforms was indicated, the sample was transferred onto MacConkey Agar for further evaluation on API20E for species identification. Approximately 15 mL of sample was added to each of the BART™ biodetector tubes (for full description, see Appendix A) and incubated at room

Table 3.2
Summary of the field parameters and data from samples taken during Phase I for water utility "A"

Water utility "A" wells	PW_5N	PW_113	PW_103	PW_136	PW_5S	PW_138
*Legionella**						
Biofilm (culture)	Positive	Negative	Negative	Negative	NA	NA
Water (culture)	Positive	Negative	Negative	Negative	Negative	Negative
Temperature (°C)*	25.4	25.5	24.1	25.2	24.6	24
Conductivity (μmho/cm)*	463	428	NA	NA	471	417
Turbidity (NTU)*	0.13	0.16	0.17	0.18	0.11	0.18
Dissolved O_2 (mg/L)*	0.6	1.04	0.65	0.81	0.7	0.48
pH (water)*	7.63	7.52	7.61	7.5	7.17	7.5
SO_4 (mg/L)	7	1	6	2	6	2
TOC (mg/L)	0.5	3.3	3.6	3.9	7.2	3.7
HPC (CFU/mL)	7	3	1	1	58	19
E.coli (CFU/mL)	Abs	Abs	Abs	Abs	Abs	Abs
Total coliforms (CFU/mL)	Abs	Abs	Abs	Abs	Abs	Abs
NH_3 (mg/L)	0.32	0.37	0.27	0.34	0.38	0.28
NO_2 (mg/L)	0.02	0.02	0.02	0.02	0.02	0.02
NO_3 (mg/L)	0.02	0.02	0.02	0.02	0.02	0.02
Viable biomass content†	1.08 10^8 cells/mL filtered (biofilm)	2.56 10^6 cells/mL filtered (biofilm)	1.66 10^5 cells/mL filtered (biofilm)	2.47 10^5 cells/mL filtered (biofilm)	9.05 10^6 cells/mL filtered (biofilm)	1.11 10^8 cells/mL filtered (biofilm)
Relatedness of microbial communities in the sample†	Relatively simple community dominated by Gram negative bacteria	Dominated by Gram negative bacteria	Dominated by Gram negative bacteria	Dominated by Gram negative bacteria	Relatively diverse community dominated by Gram negative bacteria	Relatively diverse community dominated by Gram negative bacteria
Physiological adaptations and growth phase of the bacteria in the community†	No signs of decreased membrane permeability Stationary phase	Signs of decreased membrane permeability Stationary phase	No detectable markers for membrane permeability. Late stationary phase	No detectable markers for growth phase and membrane permeability	Signs of decreased membrane permeability Stationary phase	Signs of decreased membrane permeability Log phase of growth
DGGE (dominant(s) species)†	*Methylo-microbium pelagicum* (type I methanotroph)	*Desulfosarcina* sp. *Methylophaga* sp. Thiotrix ramose	*Desulfobacteri-um indolicm; Desulfobulbus rhabdomi-formis;* Thiotrix ramose	*Desulfobacter-ium indolicum* dominant	*Desulfobac-terium* sp. *Actinonyces* sp.	Magnetobacteri um bavaricum *Desulfobacter-ium* sp. *Actinomyces* sp.

* In house determination or field measurements provided by water utility "A."
† When available.
DGGE, Denaturing Gradient Gel Electrophoresis – commissioned, Microbial Insights (MI).
PLFA, Determination of the phospholipid ester-linked fatty acids – commissioned, MI.
NA, not available.
Abs, absence.

temperature. The tubes were checked daily for one month for indicator reactions such as color changes, turbidity, sediment, precipitation, and colored formations such as rings and bubbles.

Overall, the BART™ results clearly indicate that none of the wells of water utility "A" had a biofouling problem. During 1991-1992, the predominant isolated species was *Citrobacter freundii*, and in 1994 *Enterobacter cloacae* was predominant (no identifications were performed in 1993).

Water Utility B

The procedure for biofilm sampling was different at Water Utility B. Biofilm and sediments were obtained from the water line. A length of PVC pipe (5-cm diameter) was cut off and a brush was used to collect biofilm and sediments along the pipe. Water was collected in a 4 liter bottle, as was done for Water Utility A. Field parameters measured for all the samples are recorded in Table 3.3. The biofilm collected varied in appearance depending on the well sampled. Many of the scrapings were black with black sediments present from the pipe although reddish brown scrapings and brown sediments were also observed. Water collected varied from colorless and odorless to water that had a yellow coloration and a rotten egg odor. Some samples, when filtered, contained yellow or orange particulates that collected on the filters.

CULTIVATION RESULTS AND IDENTIFICATION OF THE ISOLATES

Twenty-four samples from 12 separate wells (6 from water utility "A" and 6 from water utility "B") were received and processed. Of these, 22 tested positive for the presence of *Legionella* in the groundwater and/or the biofilm derived from the groundwater. As can be seen from the data summarized in Table 3.4, legionellae were more frequently isolated from biofilms (35 of 61 positive plates) than well water (26 of 61 positive plates).

Media that were more frequently associated with *Legionella* recovery were CCVC (37%, 23 of 61 positive plates) then BCYE (36%, 22 of 61 positive plates) then GVPA (26%, 16 of 61 positive plates). Although *Legionella* seems to be isolated more frequently from CCVC, the differences between the media are not statistically significant. Both temperatures were associated with *Legionella* recovery (35°C: 9 of 12 positive samples; 30°C: 10 of 12 positive samples). However, it should be mentioned here that some strains that grew exclusively at 30°C have been isolated, thus justifying our choice to also incubate inoculated plates at 30°C. The treatment most often associated with *Legionella* recovery was Acid Plating (24 of 61 positive plates), then Direct Plating (20 of 61 positive plates) and Heat Treatment (17 of 61 positive plates) was the least effective.

Of the 10 positive water samples, 10 were positive after inoculation without enrichment, 3 were also positive after enrichment. Of the 9 positive biofilm samples, 8 were positive after inoculation without enrichment, 1 was positive after enrichment. It is interesting to note that enrichment did not enhance the isolation of *Legionella*. In some cases, potential inhibition might have occurred during the incubation (predation and or antagonistic effects due to bacteriocinogenic strains?). Also, the number of positive *Legionella* samples from water was greater than the number of positive biofilm samples. Inhibition may also have played a role in these results since biofilm contains larger numbers of other heterotrophic bacteria. Plates are also readily overgrown, and then visual detection of suspected *Legionella* colonies becomes very difficult or impossible.

Table 3.3
Summary of field parameters and data from samples taken during Phase I for water utility "B"

		Well designation for water utility "B"					
Parameters analysed	Units	CB#1	CB#13	NWH#3	NWH#6	CC#1	CC#6
Alkalinity	mg/L as $CaCO_3$	195	177	192	201	200	215
Aluminium	mg/L	0	0	<0.003	<0.003	0	<0.003
Antimony	mg/L	<0.003	<0.003	<0.003	<0.003	<0.003	<0.003
Arsenic	mg/L	BDL	0.002	0.001	BDL	<0.001	BDL
Barium	mg/L	0.01	0.01	0.01	0.01	0.008	0.02
Beryllium	µg/L	<0.07	<0.07	<0.07	<0.07	<0.07	<0.07
Cadmium	µg/L	0.2	0.1	0.2	0.3	0.2	0.1
Calcium hardness	mg/L as $CaCO_3$	185	170	185	180	160	162
Chloride by I.C.	mg/L as Cl	6.9	5.7	9.9	8.9	8.1	7.5
Chromium	mg/L	BDL	BDL	BDL	BDL	BDL	BDL
Copper	mg/L	0.02	0.015	0.0018	0.009	0.003	0.0025
HPC	CFU/mL	51	57	31	792	2	1
Lead	mg/L	BDL	BDL	BDL	BDL	BDl	BDL
Manganese	mg/L	0.01	0.003	0.004	0.005	0.009	0.004
Nickel	mg/L	<0.001	<0.001	<0.001	<0.001	0.005	<0.001
Nitrate	mg/L	BDL	BDL	BDL	BDL	BDL	BDL
Nitrite	mg/L	BDL	BDL	BDL	BDL	BDL	BDL
Ammoniacal N	mg/L	0.25	0/09	0.19	0.24	0.12	0.14
Selenium	mg/L	BDL	BDL	BDL	BDL	BDL	BDL
Silver	mg/L	BDL	BDL	BDL	BDL	BDL	BDL
Sodium	mg/L	4.5	2.5	5.5	6	4.7	5
Sulfate by I.C.	mg/L	1	1.5	1	1.7	6	12
Turbity	NTU	3	1	0.05	0.08	0.15	0.09
Zinc	mg/L	0.03	0.03	0.01	0.02	0.03	0.04
Total Coliform [MMO-MUG]*	P or A / 100 mL	A	A	A	A	A	A
TOC	mg/L	2	1.1	3	3	2	2.5
Total sulphide	mg/L H_2S	0.04	0.03	0.08	0.2	0.06	0.15
Iron by GAA	mg/L	0.81	0.25	0.017	0.03	0.04	0.03
Conductivity	µS/cm	401	350	380	410	400	450
Temperature	°C	24.5	24	25	25	24	25
pH	units	7.2	7.5	7.5	7.5	7.5	7.6

* Total coliform testing was found mainly negative year round but some contamination episodes have occurred.
BDL, Below Detection Limit.
P = present; A = absent.

Table 3.4
Cultivation results on the samples received during Phase I (water and biofilm samples)
Calculation done on the basis of the most contaminated plate

Well/Sample #	# plates overgrown (3 days post-inoculation)	*Legionella* concentrations Biofilm* CFU/L [cfu/cm^2]/*CFU/L [cfu/cm^2]*	Water CFU/L
PW_103/E/30	12	ND	280 BW
PW_103/E/35	17	5.6 x 10^4 BW	ND
PW_103/NE/30	2	1.5 x 10^3 NA	ND
PW_103/NE/35	4	ND	100 R and 100 NA
PW_5N/E/30	16	ND	6.7 x 10^3 BW
PW_5N/E/35	16	ND	6.7 x 10^3 BW
PW_5N/NE/30	2	4.2 x 10^3 R	ND
PW_5N/NE/35	6	7.9 x 10^4 R	300 BW
PW_113/E/30	4	ND	ND
PW_113/E/35	18	ND	ND
PW_113/NE/30	6	ND	ND
PW_113/NE/35	none	500 R	ND
PW_136/E/30	4	ND	ND
PW_136/E/35	17	ND	ND
PW_136/NE/30	none	ND	ND
PW_136/NE/35	4	ND	ND
PW_138/E/30	10	ND	ND
PW_138/E/35	17	ND	ND
PW_138/NE/30	7	ND	100 NA
PW_138/NE/35	6	ND	ND
PW_5S/E/30	4	ND	ND
PW_5S/E/35	14	ND	ND
PW_5S/NE/30	none	ND	100 NA
PW_5S/NE/35	1	ND	ND
CB1/E/30	11	ND	Not detected
CB1/E/35	19	ND	Not detected
CB1/NE/30	6	100 BW [0.25 cfu/cm^2]and 1.4 x 10^3 NA [3 cfu/cm^2]/*950 BW [2 cfu/cm^2]*	100 BW
CB1/NE/35	4	ND	200 BW + 200 NA
CB13/E/30	12	ND	600 NA
CB13/E/35	10	ND	ND
CB13/NE/30	4	1.4 x 10^3 NA [3 cfu/cm^2]	100 BW
CB13/NE/35	3	ND	100 BW

(continued)

Table 3.4 (Continued)

Well/Sample #	# plates over-grown (3 days post-inoculation)	*Legionella* concentrations: Biofilm* CFU/L [cfu/cm^2]/*CFU/L [cfu/cm^2]*	*Legionella* concentrations: Water CFU/L
CC6/E/30	15	ND	ND
CC6/E/35	21	ND	ND
CC6/NE/30	none	1.0 x 10^3 NA [2 cfu/cm^2]	200 NA
CC6/NE/35	11	2.8 x 10^3 NA [6 cfu/cm^2]	400 NA
NWH3/E/30	14	ND	ND
NWH3/E/35	17	ND	ND
NWH3/NE/30	5	5 x 10^3 NA [21 cfu/cm^2]/ *3.4 x 10^3 NA [15 cfu/cm^2]*	ND
NWH3/NE/35	3	*9 x 10^3 BW [28 cfu/cm^2]/* *4.6 x 10^3 BW [20 cfu/cm^2] &* *6.7 x 10^3 NA [38 cfu/cm^2]/* *4.8 x 10^3 NA [20 cfu/cm^2]&* *235 R [2 cfu/ cm^2]/1.1 x 10^3 R [5 cfu/cm^2]*	280 NA
NWH6/E/30	9	ND	ND
NWH6/E/35	11	ND	ND
NWH6/NE/30	11	ND	800 NA
NWH6/NE/35	12	ND	200 BW & 200 NA & 280 R
CC1/E/30	21	ND	ND
CC1/E/35	18	ND	ND
CC1/NE/30	6	8.4 x 10^4 NA [84 cfu/cm^2]/ *4.5 x 10^4 NA [144 cfu/cm^2]*	100 NA
CC1/NE/35	8	2.6 x 10^4 NA [264 CFU/cm^2]	ND

* NOTE: Biofilm sample results, where possible, have been calculated based on both the volume of fluid into which the biofilm was scraped (CFU/mL) and the surface areas of the pipe sections from which the scraping was done (CFU/cm^2). This can only be done for the pipe sections sent from Water Utility B, because the scraping of Water Utility A biofilms was performed on site at the utility. Discussion of the calculation methodologies was included earlier. The results expressed in regular font are those obtained using the highest plate count whereas those in italics are a composite from all plates where it is possible for these calculations to be done.

PW, CB, CC and NWH are the wells' designation.
/NE, not enriched (not pre-incubated sample).
/E, enriched (pre-incubated sample).
/30 or /35 indicates temperature in °C.
ND = not detected.
BW (blue white), R (red) and NA (not autofluorescent) indicate the colour and presence of any autofluorescence seen.

Table 3.5
Results summary for semi-nested PCR on samples analyzed during Phase I from water utility "A"

Sample	First step 16S semi-nested PCR assay	Second step 16S semi-nested PCR assay	PCR results after addition of DNA (when necessary)*	Inhibition of PCR in the corresponding samples
PW_5N W (NE)	-	-	+	No
PW_5N B (NE)	-	-	+	No
PW_5N W (E)	-	+ (BSA)	ND	ND
PW_5N B (E)	-	+ (BSA)	ND	ND
PW_113 W (NE)	-	-	+	No
PW_113 B (NE)	-	-	+/-	Yes
PW_113 W (E)	-	-	+	No
PW_113 B (E)	-	-	+	No
PW_103 W (NE)	-	-	+	No
PW_103 B (NE)	-	-	+	No
PW_103 W (E)	-	-	+	No
PW_103 B (E)	-	-	+	No
PW_136 W (NE)	-	-	+	No
PW_136 B (NE)	-	-	+	No
PW_136 W (E)	-	-	+	No
PW_136 B (E)	-	-	+	No
PW_138 W (NE)	-	-	-	Yes
PW_138 B (NE)	-	-	+/-	Yes
PW_138 W (E)	-	-	-	Yes
PW_138 B (E)	-	+ (BSA)	+	No
PW_5S W (NE)	-	-	+	No
PW_5S B (NE)	-	-	+/-	Yes
PW_5S W (E)	-	+ (BSA)	+	No
PW_5S B (E)	-	+ (BSA)	+	No

* DNA added (2 µl - 400 ng - of extracted DNA from *L. steigerwaltii*) to check for the presence of potential PCR inhibitors in the remaining negative samples (Semi-nested PCR). In this case only the first step of the semi-nested PCR has been done.
ND, not done.
W = Water (heat-thawing lysis); B = Biofilm (heat-thawing + buffer lysis).
E = enriched; NE = not-enriched.
5 µl per reaction (first step 16S semi-nested PCR).
1 µl of PCR product from step 1 for second step 16S semi-nested PCR.
BSA (addition of BSA in the first step semi-nested PCR assay).

Although the number of wells sampled was too small to have generated significant results, their processing has produced definite indications with regards to the best treatment(s) to enhance the recovery of *Legionella* from groundwaters, while suppressing the growth of other heterotrophic microflora. The enrichment procedure, which was intended to enhance cultivation was not used in Phase II because it produced many more overgrown plates and a reduced number of culture-positive samples. Such overgrown plates made it very difficult or impossible to visually detect legionellae since the the putative colonies are selected based on visual appearance.

Detailed records were kept for all cultivation, on the examination of the culture plates, sub-cultivation of suspected colonies, and confirmation of their identity. However, it is not possible to include those details here. Strain identification was done on both fluorescent and non-fluorescent suspect colonies. Particular attention was however paid to non-fluorescent colonies, and partial 16S rDNA sequencing gave interesting results with regards to species diversity. For instance, it would be interesting to confirm the identity of some isolated strains (e.g., E103 from PW_103 and ACB130 from CB#1). E103 is related (by partial 16S rDNA sequencing) to *L. geestiana* (percentage of homologies for the fragment sequenced is 97%), a very rare species for which only one strain has been described so far. ACB130 grew only at 30°C and is closely related (by partial 16S rDNA sequencing) to *L. rubrilucens* (percentage of homologies for the fragment sequenced is 96%) and two LLAPs strains (LLAP-9 and LLAP_6) (percentage of homologies for the fragment sequenced is 96%) as well as *L. lytica* (percentage of homologies for the fragment sequenced is 96%), a recently described uncultivable species (Hookey et al., 1996) formerly known as *Sarcobium lyticum* (Drozanski 1991)). Interestingly, ACB130 does not exhibit a typical red autofluorescence (which is the case for *L. rubrilucens*). Other isolated strains might also be interesting for further studies.

Identification results for Phase I are summarized in Chapter 4 (Table 4.3), along with those for Phase II and tentative clustering between isolates is shown in Appendix F. However, it is important to note that determination of strain identity from 16S rDNA sequencing is not the definitive method for identifying *Legionella* species and the strain identities obtained with partial 16S rDNA sequencing should be considered with caution, especially when a definitive bacterial identification is needed. For example, it seems unlikely that very rare species, only encountered once since their first description, should be found during this study. On the other hand DGGE analysis of samples from the same utility also suggested this species was present.

Summary for Phase I cultivation results:
Number of water samples culture-positive: 10 of 12
Number of biofilm samples culture-positive: 9 of 12

Semi-nested PCR Assay

Results for *Legionella* detection using the semi-nested procedure have been summarized for Phase I (Tables 3.5 & 3.6). Of the 24 samples tested, 7 were PCR positive. All positive samples were obtained after incubation (pre-enrichment) of the samples for one week at 35°C. This enrichment procedure likely increased the number of *Legionella* in the samples so that detection became possible by PCR. However, as mentioned before, the enrichment procedure does not increase the level of detection by cultivation. It is to be underlined that pre-enrichment of samples leads frequently to an overgrowth of the plates so that a visual detection of the typical *Legionella* colony is no longer feasible. For Phase II, enrichment was used only prior to PCR detection.

Table 3.6
Results summary for semi-nested PCR on samples analyzed during Phase I from water utility "B"

Well-Sample #	First step 16S semi-nested PCR assay	Second step 16S semi-nested PCR assay	PCR results after addition of DNA (when necessary)*	Inhibition of PCR in the corresponding samples
CB #1 W (NE)	-	-	-	Yes
CB #1 B (NE)	-	-	-	Yes
CB #1 W (E)	-	-	+	No
CB #1 B (E)	-	-	-	Yes
CB #13 W (NE)	-	-	+	No
CB#13 B (NE)	-	-	+	No
CB #13 W (E)	+ (BSA)	+	+	No
CB #13 B (E)	-	+	-	Yes
CC #6 W (NE)	-	-	+	No
CC #6 B (NE)	-	-	-	Yes
CC #6 W (E)	-	-	+	No
CC #6 B (E)	-	-	-	Yes
NWH #3 W (NE)	-	-	+	No
NWH #3 B (NE)	-	-	-	Yes
NWH #3 W (E)	-	-	+	No
NWH #3 B (E)	-	-	-	Yes
NWH #6 W (NE)	-	-	+	No
NWH #6 B (NE)	-	-	-	Yes
NWH #6 W (E)	-	-	+	No
NWH #6 B (E)	-	-	-	Yes
CC #1 W (NE)	-	-	+	No
CC #1 B (NE)	-	-	-	Yes
CC #1 W (E)	-	-	+	No
CC #1 B (E)	-	-	-	Yes

* DNA added (2 µL - 400 ng - of extracted DNA from *L. steigerwaltii*) to check for the presence of potential PCR inhibitors in the remaining negative samples (Semi-nested PCR). In this case only the first step of the semi-nested PCR has been done.
W = Water (heat-thawing lysis); B = Biofilm (heat-thawing + buffer lysis).
E = enriched; NE = not enriched.
5 µL per reaction (first step 16S semi-nested PCR).
1 µL of PCR product from step 1 for second step 16S semi-nested PCR.
BSA (addition of BSA in the first step semi-nested PCR assay).

Six biofilms and two water samples contained PCR inhibitors possibly leading to false-negative results. It is not known why these biofilm samples (different wells) contained PCR inhibitors. It is likely that one or more common substance(s) might be responsible for such an inhibition. Inhibitory substances detected in the enriched biofilm from CB #13 were sufficiently diluted by the semi-nested PCR procedure to allow an efficient amplification. However, in other cases such as NWH3, CC6 and CC1, the dilution of the second step of the PCR was not sufficient to decrease the inhibitory effect of the extract. On the basis of the results observed by cultivation, all these biofilm samples should be considered as false-negative samples. No attempts were made in this study to remove the PCR inhibitors prior to testing because of potential difficulties in also removing Legionellae when present in small numbers.

Summary for Phase I PCR results:

Number of water samples PCR-positive: 3 of 12 processed

Number of biofilm samples PCR-positive: 4 of 12 processed

DISCUSSION AND CONCLUSIONS

The purpose of the Pilot Study (Phase I) was to evaluate whether groundwater systems were a potential niche for *Legionella* because it had not been unambiguously proven if these ecosystems harbored bacteria primarily known to be associated with surface water environments. In addition, the Pilot Study was undertaken to evaluate if the methodological approaches chosen by the research team were appropriate to investigate the occurrence of *Legionella* in such environments. The sample sites selected had elevated groundwater temperature due to their geographical location and were not believed to be under the influence of surface water. Clearly the pilot study demonstrated the success of the methodological approach since legionellae were frequently isolated. Overall, *Legionella* (no specific identification of strains was initially requested) were detected by cultivation in 19 out of the 24 samples tested (each well was tested for *Legionella* in both water and biofilm samples) with concentrations ranging from 10^2 to 8×10^4 CFU/L. PCR was positive for 7 out of the 24 samples tested. *Legionella* were shown to colonize the two warm (average water temperature for the samples tested is 25°C) groundwater systems tested. On the basis of this observation it can be anticipated that *Legionella* may be part of the groundwater microflora and that they cannot be, strictly speaking, considered only as a surface water pathogen. The wells investigated during the Pilot study were from two separate wellfields. Although they may not be considered as representative of the global situation within U.S. warm groundwater systems, it is possible that they are predictive of similar results elsewhere. The fuller assessment study in Phase II aimed to verify or refute this hypothesis.

An alternative hypothesis, which cannot be completely discounted, is that the wells tested, even if very deep, showed some, potentially remote, influence of surface water, and that they had become colonized with legionellae as a result of this association.

PERSPECTIVE AND CONSEQUENCES FOR PHASE II

Phase II of this study was designed to conduct a fuller assessment of the overall presence of legionellae in ground waters throughout North America. It was decided to sample both cold and warm groundwater systems - if the cold waters proved negative for legionellae, then the original hypothesis regarding temperature would be further supported. If, on the other hand, legionellae were found even in 'cold' (< 15°C) groundwaters, then it would truly emphasize their ubiquitous nature in water environments.

Survey Implementation

By agreement with members of the AwwaRF Project Advisory Committee (PAC), for Phase II, one water utility (referred to as "Water Utility A" during Phase I) was selected for a one-year survey to evaluate if potential variations in the number and the species distribution of *Legionella* could be observed (since it is believed that microbial contamination may be transient). For this purpose, two wells (PW_103 and PW_136) tested during Phase I were regularly sampled (once a month) for *Legionella* analysis. PW_103 was found positive for *Legionella* during Phase I and PW_136 was negative. It was decided, in conjunction with PAC members, to select these two wells for a survey. The other utilities participating in Phase II were obtained by a widespread electronic and telephone solicitation of interest. Other utilities participated on a single event basis where water and biofilm samples were sent from one or two individual wells.

Data Collection

Data collection by participating water utilities was strongly encouraged by the research team. It was quite successful in Phase I, but not all water utilities participating in the study were able to provide the details of all the environmental parameters of interest (see also Phase II of the study). Thus, the potential to analyze for correlations that may indicate the presence of *Legionella* in the samples was restricted. Sample analyses for bacterial indicators of fecal pollution were always negative, whether tested in house by the utilities, or in the CREM laboratories. It is clear therefore that legionellae can be found even in the absence of conventional bacterial indicators of fecal pollution, and this was an anticipated outcome.

Modifications of the Technical Approach

Cultivation

As shown in our results from Phase I, preincubation of the samples did not increase the recovery of *Legionella* by cultivation. This procedure, which was beneficial to release potential intracellular *Legionella* (amoebae) within the samples, was also shown to increase the level of heterotrophic associated microflora. These contaminating organisms overgrow *Legionella* on the media despite the use of several decontamination procedures. On the other hand, *Legionella* was better detected by PCR when pre-incubation of the samples was conducted. Therefore, it was decided for Phase II to not use pre-incubation for cultivation but that this "pre-enrichment" stage would still be done to favor PCR detection. Therefore, cultivation in Phase II would be carried out only from non-enriched samples as opposed to Phase I when both non-enriched and enriched samples were processed.

As we hypothesized very early when drafting our proposal, the incubation temperature may be a critical parameter when trying to isolate *Legionella* from environmental samples. Isolation of legionellae from clinical specimens is normally performed at 35°C and this has become the conventional temperature used for all *Legionella* isolations. However, when dealing with environmental samples, a lower incubation temperature may be less stressful for the bacterium and could ensure a better recovery on the plates. During Phase I, all the inoculated plates were incubated at both 35 and 30°C to ensure the best possible recovery of the bacteria. It was decided to continue this procedure in combination with all the media and the decontamination steps during Phase II without any changes in the protocols. The only change was to eliminate anisomycin as a supplement for GVP and to use two GVP plates instead of one during isolation.

During Phase I, all cultivation results were expressed in CFU/L, and for Phase II it was decided that results from biofilm cultivation will be expressed in CFU/cm^2 for a better representation of the presence of legionellae in surface biofilms. Also, as discussed earlier, a change was made in the calculation strategy and the *Legionella* number estimates were based on the highest counts from individual plates rather than on the mean values which were rather distorted by overgrowth on many plates.

PCR Assay

The sensitivity of the PCR was surprisingly poor. This was despite using a semi-nested procedure combining a two-step amplification process. It was anticipated that the semi-nested procedure would reduce the effect of potential PCR inhibitors by a dilution mechanism between the two amplification steps. In view of that, during Phase II, an improved DNA extraction method (a commercially available DNA extraction kit: Dneasy extraction kit, Qiagen) was used to try to increase the global sensitivity of the assay (also possibly reducing the effect of the PCR inhibitors). It should be noted that the majority of the samples tested had low *Legionella* concentrations. These low concentrations were at or below the detection threshold of the PCR assays.

Immunomagnetic Separation (IMS)

IMS is a selective technique based upon the ability of a specific antibody coated onto magnetic beads to bind the study pathogen which can then be removed from a complex mixture simply by use of attracting the beads to a permanent magnet. The approach is very attractive because of its apparent simplicity and ability to simplify the isolation procedure. It was planned to add development of an IMS procedure for legionellae as a part of the Phase II study, with the expectation that if successful it would greatly simplify the selective isolation of *Legionella* from highly contaminated groundwater samples. It was expected that IMS in conjunction with cultivation may reduce the need for decontamination treatments and potentially reduce the number of plates needed. Also, combining IMS with PCR could improve the detection of *Legionella* in various samples.

Quality Assurance and Quality Control

The QA/QC plans elaborated during Phase I were found to be satisfactory and were applied unchanged during Phase II of the study.

CHAPTER 4
RESULTS: PHASE II, ASSESSMENT STUDY (PROJECT NUMBER 2657)

In order not to create any confusion with water utilities that collaborated during Phase I, the name "Water Utility B" has been dropped for this section of the report. Therefore, in addition to Water Utility A, the participating water utilities are referred to as "Water Utility C," "Water Utility D," etc.

DESCRIPTION OF PARTICIPATING WATER UTILITIES

During Phase II, one Water Utility (referred to as "Water Utility A" during Phase I) was selected for a one-year survey to evaluate if potential variations in the number and the species distribution of *Legionella* could be observed. It was decided to select two wells (PW_103 and PW_136) that had been tested monthly for *Legionella* during Phase I. PW_103 was found positive for *Legionella* during Phase I and PW_136 was negative. The participating water utilities were free to select the type of pipe material(s) best suited to their distribution system. As an example, "Water Utility A" used 2-inch Blue Brutte John Mansfield pipe. All other water utilities provided samples only once during Phase II. Descriptions of the samples provided by the participating utilities are found in Appendix D, and field data for the samples in Appendix E.

Although the research team had decided to focus on raw water or biofilm samples directly from the wellhead, this was not always possible when dealing with utilities providing single samples and unable to directly access this area with ease. Therefore on occasion, when investigating the presence of *Legionella*, it was necessary to use water and/or biofilm samples taken from the distribution system but as close to the wellhead as possible to avoid any potential for contamination from the distribution system itself.

Although the research team was able to process many additional samples for the survey, recruitment of participating water utilities was not simple. Different sampling strategies were proposed to water utilities and although the research team recognizes that collecting data pertaining to the samples and the wells investigated may have been time consuming, it was surprised that the study did not attract more interest. In the words of one utility employee, 'it was a lot of work for a single sample,' and this utility claimed that it would have been easier to set up and supply samples regularly from a single well. Unfortunately, taking regular samples from several municipalities would not provide the survey component that the study requested, and doing so from a large number of utilities would have totally overwhelmed the capacity of the study team. Several utilities that initially expressed interest in providing samples failed to do so.

LEGIONELLA AND PROTOZOA

During Phase II, we decided to investigate, as an additional parameter and for selected samples, the presence of protozoa. Protozoa analyses were conducted on some samples of waters and biofilms to evaluate whether species historically associated with *Legionella* (for instance, *Naegleria*, *Acanthamoeba*, *Vahlkampfia*, and *Hartmanella*) could be detected in the samples. The biofilm serves as a focal point where bacterial and protozoan populations interact. The pioneering bacterial population will modify the surface conditions to enable bacterial succession to take place. The resultant biofilm may aid colonization by *Legionella* by a) supporting bacterial flora that provide essential nutrients; b) providing suitable growth conditions by removing high/inhibitory concentrations of oxygen by respiration; or c) by encouraging protozoan grazers which can act as hosts for legionellae.

It should be noted that, in addition to biofilm, the relationship between *Legionella* species and certain free-living protozoa may affect normal disinfection practice. Certain free-living protozoa permit the intracellular growth and multiplication of *Legionella*. The trophozoites of such amoebae, and *Legionella*, are normally inactivated by 1-2 ppm active chlorine, and thus the intracellular location of *Legionella* would not normally constitute a problem for infection control. Viable *Legionella*, however, can be recovered from the protozoan cysts; the latter can resist 50 ppm of chlorine and *Legionella* from protozoan cysts may demonstrate an enhanced resistance to chlorine.

Another study has shown that, in transplant-unit cold water systems (these cannot be compared with groundwater ecosystems), a significant association between the isolation of *Legionella* and protozoa known to permit the intracellular growth of *Legionella* was observed (Patterson et al. 1994; Patterson et al. 1997). In hot transplant unit water systems, an inverse association between the presence of *Legionella* and water temperature has been found. (*Legionella* was found in distribution systems with circulating water temperatures as high as 66°C. *Legionella* was also isolated from distribution systems with circulating water temperatures as low as 8°C). The complex biological relationship between *Legionella* and protozoa, which is, at least in part, temperature dependent, may be relevant in the context of groundwater ecosystems.

The hypothesis that legionellae gain entry to domestic water supplies by continual seeding from ruptured protozoa is further supported by the finding of an association between *Legionella* isolation from cold water distribution systems and the isolation of protozoa known to permit the intracellular growth of *Legionella* (Patterson et al. 1994; Patterson et al. 1997). During this study, *L. pneumophila* was isolated most commonly from hot-water supplies and the blue-white autofluorescent *Legionella* species most commonly from cold-water supplies, but these differences were not significant statistically. Protozoa were isolated more commonly from hot water supplies (Patterson et al. 1994; Patterson et al. 1997).

LEGIONELLA DATA

Cultivation for *Legionella*

Cultivation results are presented below by utility in Tables 4.1 and 4.2. During this phase of the study, optimization of the IMS procedures was being undertaken and IMS was included for some samples. When possible, the IMS-cultivation assay was conducted in parallel with the conventional cultivation assay and results are given below. Results of analyses for total and fecal coliform, and fecal strep (not shown) were all negative, whether data was conducted in house or in the CREM laboratory. Heterotrophic plate counts varied from less than 10 to 10^4 to 10^5 CFU/100 mL (results not shown).

Table 4.1
Legionella cultivation results for Water Utility A (Florida)

Well ID	Date	Water (CFU/L)	Isolation temp. (°C)	Strain identification	Biofilm (CFU/cm^2)	Isolation temp. (°C)	Strain identification
103-1	Mar-Apr 2000	280	30	*L. geestiana*	None	N/A	N/A
136-1		None	N/A	N/A	None	N/A	N/A
103-2	Apr-May 2000	None	N/A	N/A	1	30	No sequence
136-2		None	N//A	N/A	None	N/A	N/A
103-3	May-Jun 2000	None	N/A	N/A	None	N/A	N/A
136-3		None	N/A	N/A	None	N/A	N/A
103-4	Jul-Aug 2000	None	N/A	N/A	None	N/A	N/A
138-1		431	30	No sequence	None	N/A	N/A
103-5	Aug-Sep 2000	None	N/A	N/A	None	N/A	N/A
138-2		413	30	No sequence	None	N/A	N/A
103-6	Oct-Nov 2000	None	N/A	N/A	None	N/A	N/A
138-3		None	N/A	N/A	None	N/A	N/A
103-7	Dec-Jan 2001	None	N/A	N/A	None	N/A	N/A
138-4		4600	30	*L. parisiensis*	None	N/A	N/A
103-8	Jan-Feb 2001	None	N/A	N/A	None	N/A	N/A
138-5		206	30	*L. parisiensis*	None	N/A	N/A
103-9	Feb-Mar 2001	150	35	*L. erythra*	None	N/A	N/A
		100	35	*L. gresilensis*			
		100	30	*L. erythra*			
		150	30	*L. gresilensis*			
138-6		206	35	*L. santicrucis*	None	N/A	N/A
		100	35	*L. gresilensis*			
		100	30	*L. gresilensis*			
		100	30	*L. waltersii*			
103-10	Apr-May 2001	100	35	*L. dumoffii*	None	N/A	N/A
138-7		413	30	LLAP-2	None	N/A	N/A
103-11	May-Jun 2001	N/R	N/A	N/A	6.4	?	*L. dumoffii*
					15.9	?	*L. erythra*
					13.1	?	*L. quateirensis*
138-8		None	N/A	N/A	None	N/A	N/A
103-12	July 2001	N/R	N/A	N/A	14.4	?	*L. quateirensis*
					29.6	?	*L. erythra*
138-9		619	35	*L.quateirensis*	None	N/A	N/A
103-13	Sep 2001	N/R	N/A	N/A	None	N/A	N/A
116-1		100	35	*L. erythra*	None	N/A	N/A
138-10		100	35	*L. feelei*	None	N/A	N/A
116-2	Oct 2001	50	35	*L. dumoffii*	None	N/A	N/A
138-11		100	35	*L. erythra*	None	N/A	N/A
		25	30	LLAP-2			
KWII RW	Nov 2000	N/R	N/A	N/A	206	30	Blue-white
					100	35	autofluorescent
KWII IFS	Nov 2000	N/R	N/A	N/A	None	N/A	N/A
05B	Apr 2001	N/R	N/A	N/A	206	30	LLAP 10
					500	35	LLAP 10
15A	Apr 2001	N/R	N/A	N/A	None	N/A	N/A
21A	Apr 2001	N/R	N/A	N/A	None	N/A	N/A
32G	Apr 2001	N/R	N/A	N/A	206	30	*L. anisa*
					200	35	LLAP 10
72F	Apr 2001	N/R	N/A	N/A	None	N/A	N/A
94A	Apr 2001	N/R	N/A	N/A	100	35	LLAP 10

N/R indicates sample not received; N/A indicates not applicable

Water Utility A

In summary, PW_103, which was the positive well (during Phase I) selected for survey, was found positive for *Legionella* when the first samples were obtained at the beginning of Phase II. Significant differences in terms of level of contamination and species were found suggesting possible variations from time to time in the same well. Changes in wells sampled as a result of unforeseen events are described in Appendix A.

After a period when *Legionella* were undetected in the samples received (May 2000 to February 2001), PW_103 was found positive again for the last samples tested which were sometimes contaminated with at least two different species (low concentrations). Interestingly, different species were sometimes recovered in two successive samples.

Following the general observation that the frequency of legionellosis cases over a year is known to follow a seasonal distribution with higher number of cases during spring, it is interesting to speculate whether, when looking at the frequency of *Legionella* isolation in this well, legionellosis cases were correlated with the higher prevalence of *Legionella* at the same period of the year. However, this distribution of *Legionella* was not verified in the other well (PW_138) surveyed (see below). This well was nevertheless regularly contaminated with *Legionella*. Overall, this well (both water and biofilm samples) was found positive for *Legionella* with sometimes high variations in bacterial concentrations from month to month and with a remarkable diversity of species recovered.

It is interesting to note that so far LLAP strains are mostly considered as 'not yet cultivable' species (under standard conditions). It is surprising to be able to isolate such a species on conventional BCYEα media without special supplements. Therefore, it can be speculated that, this species might be better able to grow at 30°C than at 35°C (the usual temperature at which all laboratories are incubating their media). The other possibility is that the identification of the isolate is incorrect however the percentages of identity are such that a misidentification is improbable (98% identities with LLAP-2, LLAP-9, LLAP-6 and with *L. lytica* (L2 strain), and 97% identities with LLAP-7).

Some ciliates, possibly *Tetrahymena* or more likely *Microthorax*, were also regularly found in PW_138 water samples suggesting that there might be a protozoan population that supports *Legionella* growth within this well.

Well #103 tested only slightly positive for *Legionella* during Phase I, but was found surprisingly positive for *Legionella* in almost all the samples tested. Overall, a high diversity in the *Legionella* species was recovered (except November-December 2000 and May-June 2001 samples). The February-March 2001 sample was contaminated with different species of *Legionella* (as stated above).

The lack of *Legionella* recovery for the May-June 2000 sampling events is likely to be due to the rehabilitation procedure that was conducted on this well at this period. This was not a true rehabilitation program, it was more of a well maintenance and testing program; the procedure involved pulling the well pump, scraping the biofilm off the well casing and chlorinating the well. In contrast, when a well is rehabilitated at Water Utility A, any or all of the following procedures are used: The open hole, down deep in the well is acidified, liquid CO_2 may be added to the hole, and/or an air hose may be lowered into the hole to break the biofilm in the well loose. Then a submersible pump will pump out the debris. The well is chlorinated, reassembled and pumped until it passes the bacteriological tests. After these sampling events, the well almost constantly tested positive for *Legionella* despite the rehabilitation procedure.

Some isolates were frequently recovered that possibly correlated to LLAPs and only grew at 30°C. Other atypical isolates were suspected although a definitive identification was not achieved.

The KII RW biofilm samples from these same wells had a very heavy growth probably due to the large amount of biofilm material extracted. Decontamination procedures known to decrease the number of heterotrophic associated-bacteria were not efficient. All the plates were overgrown. It was decided to dilute (1/10) the concentrate and to re-do the testing. After this dilution, some bluish-white auto-fluorescent colonies were found but due to heavy contamination these colonies had to be isolated from the bacteria that were overgrowing them. This was done by streaking the colonies out onto BCYEα media.

Water Utilities C-Q

The samples obtained from Water Utility C (Arizona) are contaminated with several species of *Legionella* (Table 4.2). The more contaminated sample is the water meter where the sample was scraped from a 10 year-old biofilm.

Water utility D (Texas) samples were obtained from a booster station/reservoir fed by 34 wells. The well/reservoir feeds the distribution system (samples thiosulfate-preserved). Although some strains had the morphological appearance of *Legionella* and had the cysteine requirement for their growth, no PCR product was obtained with our *Legionella* PCR assay. Therefore, these strains were not considered as *Legionella* strains. Although the samples were culture-negative, the fact that the samples were found positive by PCR (see below) indicates that uncultivable or poorly cultivable forms might be present. These non-cultivable forms could be affected by the chlorine residual (the samples were obtained very close to a booster station). It is also possible that our PCR assay had detected naked or released DNA from dead cells.

Samples obtained from Water Utility F (Texas), which is known to have extremely warm groundwaters (temperatures > 40°C), have been found positive for *Legionella* (Table 4.2). Due to the high temperature, it was found that the samples had few other types of contaminating bacteria suggesting that a thermal selection occurs in the wells tested.

Samples obtained from Illinois (Water Utility E, Water Utility G and Water Utility H) did not contain cultivable forms of *Legionella* at the time of the analysis. The water sample from the first utility was nevertheless found positive by PCR testing. However, the fact that the water samples received were considered "finished water," and that the biofilm samples were scraped from meters located very close to booster stations, they may not be representative of the real situation in the production wells (not accessible), or in the distribution system away from the booster stations.

Legionellae were isolated from Water Utilities J (Michigan), O (Missouri), P (Nebraska) and Q (Florida), but not from Water Utilities I (Missouri), M (Illinois) or N (New Brunswick). The last samples received from two utilities (New Jersey, and Ontario) are positive by cultivation suggesting that *Legionella* can also colonize/survive in cold groundwater systems (an expected result).

It is important to remember that the results from all utilities other than Utility A represent only a simple snapshot of a small sample of water on a single occasion. In some cases utilities submitted samples from a single well and in other cases two wells were sampled.

In summary, *Legionella* colonies were detected on all media used in this study but 47% of the positive plates used CCVC medium. GVP gave 30% of the positive plates whereas BCYEα

Table 4.2
***Legionella* cultivation results for water utilities C to Q**

Water utility	Well ID	Date	Water CFU/L	Incubation temperature (°C)	Strain identification	Biofilm source	Biofilm CFU/cm^2 or as stated	Incubation temperature (°C)	Strain identification
C	AZ (A)	May 2000	100	30	Not Auto Fluor. Sp.	N/R	N/A	N/A	N/A
Arizona	AZ (B)		200	30	BW species	Coupons Meter	None	N/A	N/A
	AZ (BM)		N/R	N/A	N/A		5	30	BW species
							5	30	*L. pneumophila*
							3	30	Non-related *L.* sp.
							20	35	BW species
							5	35	*L. pneumophila*
D Texas	EP	May 2000	Nonc	N/A	N/A	Meter	None	N/A	N/A
E Illinois	ST	June 2000	None	N/A	N/A	Meter	None	N/A	N/A
F	CS2	July	100	30	*L. pneumophila*	Pipe sections	10	30	*L. pneumophila*
Texas		2000	100	35	*L. pneumophila*		5	35	*L. pneumophila*
	CS5		183	30	*L. pneumophila*		5	30	*L. pneumophila*
G Illinois	PEO	August 2000	None	N/A	N/A	Water meter	None	N/A	N/A
H Illinois	PK	Sept. 2000	None	N/A	N/A	Water meter	None	N/A	N/A
I Missouri	MEX	Oct. 2000	None	N/A	N/A	Water mater	None	N/A	N/A
J	SF	Feb.	100	30	*L. donaldsonii*	Pipe sections	None	N/A	N/A
Michigan	MN	2001	None	N/A	None		None	N/A	N/A

(continued)

Table 4.2 (Continued)

Water utility	Well ID	Date	Water CFU/L	Incubation temperature (°C)	Strain identification	Biofilm source	Biofilm CFU/cm² or as stated	Incubation temperature (°C)	Strain identification
K New Jersey	NJ	April 2001	None	N/A	N/A	Meter	118	35	ATCC 700511
L Ontario	RI	April 2001	412	30	*L. quateirensis*	Scraped in water	6.2x10³ CFU/L	30	*L. quateirensis*
M Illinois	CH	May 2001	None	N/A	N/A	Water meter	None	N/A	N/A
N New Brunswick	NB3	June 2001	None	N/A	N/A	Pipe sections	None	N/A	N/A
	NB7		None	N/A	N/A		None	N/A	N/A
O Missouri	MO18	June 2001	100	35	*L. quateirensis*	Scraped in water	None	N/A	N/A
			300	30	*L. quateirensis*				
	MO34		None	N/A	N/A		None	N/A	N/A
P Nebraska	NE54	July 2001	None	N/A	N/A	Pipe sections	None	N/A	N/A
	NE66		100	30	*L. quateirensis*		14	30	*L. quateirensis*
			412	30	*L. dumoffii*				
Q Florida	FK8	July 2001	None	N/A	N/A	Not received	N/R	N/A	N/A
	FK13		800	35	*L. pneumophila*		N/R	N/A	N/A

BW = Blue-white autofluorescence; N/R = not received; N/A = not applicable.

allowed the detection of *Legionella* on 23% of the positive plates. The presence of antibiotics was therefore beneficial for detection of legionellae, presumably by suppressing some of the accompanying bacteria. More *Legionella* isolates were obtained following incubation at 30°C. Acid treatment was found to be better than heat or no treatment to recover legionellae from the samples investigated. Overall, 34% of the samples (36 out of 105) were found positive for *Legionella* by cultivation.

Summary for Phase II cultivation results:
Number of water samples culture-positive: 23 of 48 (48%)
Number of biofilm samples culture-positive: 13 of 57 (23%)

Strain Identification Results

Isolates were identified by nucleotide sequencing of a 531-basepair fragment of their 16S rRNA gene. Nucleotide sequences of the amplified fragments (first step semi-nested PCR) obtained from suspected colonies were compared with available sequences in the DDBJ, EMBL and GenBank databases with the Gapped BLAST program (Altschul et al. 1990) through the Internet (http://www.ncbi.nlm.nih.gov). Sequences were then aligned with each other, and with available 16S rDNA sequences of the recognized *Legionella* species using the Clustal w1.6 alignment computer program (Thompson, Higgins, and Gibson, 1994). Phylogenetic analyses were carried out on strains with sequences available using the PHYLIP 3.572c Phylogenetic Inference Package (Felsenstein, 1981). Unrooted trees were calculated from the distance method (DNADIST). A fragment of the 16S rDNA published sequence of *Coxiella burnetii* was used for the outgroup analysis due to its high level of homology with *Legionella* 16S rDNA sequences. Pairwise distances between all sequences were inferred under the maximum-likelihood (DNADIST) criteria. Trees were constructed by using the algorithm of Fitch and Margoliash (Fitch and Margoliash, 1967).

Some isolates appear unambiguously identified, but others remain unclustered indicating possible new species or atypical isolates and are hence unidentified. These results should be considered with caution because partial 16S rRNA gene sequencing is not the definitive tool to assign a species name to an isolate. Some isolates growing at 30°C may be unrecognized species or strains. Most of the laboratories working on *Legionella* detection do not incubate their plates at 30°C; therefore, new species could be possible from our work.

A summary of the identification done on selected strains isolated during Phases I and II is provided in Table 4.3. Dendrograms showing clustering of isolates among recognized species are shown in Appendix F (Phase I isolates in bold) and Appendix G (Phase II isolates underlined).

Table 4.3
Identification of selected *Legionella* isolates from Phases I and II
(BLAST search through the Internet, http://www.ncbi.nlm.nih.gov)

Strain denotation and corresponding water utility	Strain identification (BLAST search)
ACB1 30 (CB1)	96% identities with *L. rubrilucens*
Water Utility "B"	96% identities with *L. sp* strain LLAP-9
PHASE I isolate	96% identities with *L. rowbothamii* (formerly LLAP-6)
	96% identities with *L. lyticum* strain LLAP-3,
	96% identities with *L. lyticum* strain L2
C2TPB1 (CB1)	99% identities with *L. rubrilucens* (WA-270A-C2)
Water Utility "B"	99% identities with *L. taurinensis*
PHASE I isolate	99% identities with *L. rubrilucens* (ATCC 35304)
	98% identities with *L. erythra*
C1 (PW_5N)	99% identities with *L. dumoffii*
Water Utility "A"	98% identities with *L. gormanii*
PHASE I isolate	
C21 (PW_5N)	99% identities with *L. erythra*
Water Utility "A"	98% identities with *L. rubrilucens*
PHASE I isolate	98% identities with *L. taurinensis*
E103 (PW_103)	97% identities with *L. geestiana*
Water Utility "A"	92% identities with *L. erythra*
PHASE I isolate	92% identities with *L. quinlivanii*
	91% identities with *L. pneumophila* (E5C1)
	91% identities with *L. sp* strain LLAP-9
	91% identities with *L. sp* strain LLAP-7
	91% identities with *L. drozanskii* (formerly LLAP-1)
	91% identities with *L. lyticum* strain L2
G103 (PW_103)	99% identities with *L. erythra*
Water Utility "A"	98% identities with *L. rubrilucens*
PHASE I isolate	98% identities with *L. taurinensis*
	97% identities with *L. brunensis*
PW_113 red (PW_113)	99% identities with *L. erythra*
Water Utility "A"	98% identities with *L. rubrilucens*
PHASE I isolate	98% identities with *L. taurinensis*
K103/1 red (PW_103)	97% identities with *L. geestiana*
Water Utility "A"	92% identities with *L. drozanskii* (formerly LLAP-1)
Growth at 30°C	92% identities with *L. erythra*
PHASE II isolate	92% identities with *L. donaldsonii*
Identical to E103	92% identities with *L. pneumophila*
(Isolated from the same well)	

(continued)

Table 4.3 (Continued)

Strain denotation and corresponding water utility	Strain identification (BLAST search)
AB(AZ)/35/1	99% identities with *L. anisa*
Water Utility "C"	99% identities with *L. sp.* (strain IBV ATCC 700511)
Growth at 35°C	99% identities with *L. bozemanii*
PHASE II isolate	98% identities with *L. parisiensis*
	99% identities with *L. dumoffii*
E(AZ)/35/1	99% identities with *L. anisa*
Water Utility "C"	99% identities with *L. sp.* (strain IBV ATCC 700511)
Growth at 35°C	99% identities with *L. bozemanii*
PHASE II isolate	98% identities with *L. parisiensis*
	98% identities with *L. dumoffii*
F(AZ)/35/1	98% identities with *L. sp.* (strain IBV ATCC 700511)
Water Utility "C"	98% identities with *L. anisa*
Growth at 35°C	98% identities with *L. bozemanii*
PHASE II isolate	98% identities with *L. parisiensis*
	98% identities with *L. dumoffii*
G(AZ)/35/1	99% identities with *L. anisa*
Water Utility "C"	99% identities with *L. sp.* (strain IBV ATCC 700511)
Growth at 35°C	98% identities with *L. dumoffii*
PHASE II isolate	98% identities with *L. bozemanii*
	97% identities with *L. parisiensis*
I(AZ)/35/1	99% identities with *L. anisa*
Water Utility "C"	98% identities with *L. sp.* (strain IBV ATCC 700511)
Growth at 35°C	98% identities with *L. bozemanii*
PHASE II isolate	98% identities with *L. dumoffii*
	98% identities with *L. parisiensis*
L(AZ)/35/1	99% identities with *L. anisa*
Water Utility "C"	99% identities with *L. sp.* (strain IBV ATCC 700511)
Growth at 35°C	98% identities with *L. bozemanii*
PHASE II isolate	98% identities with *L. dumoffii*
	98% identities with *L. parisiensis*
M(AZ)/35/1	99% identities with *L. anisa*
Water Utility "C"	99% identities with *L. sp.* (strain IBV ATCC 700511)
Growth at 35°C	99% identities with *L. bozemanii*
PHASE II isolate	99% identities with *L. dumoffii*
	98% identities with *L. parisiensis*
X(AZ)/30/1	96% identities with *L. rubrilucens*
Water Utility "C"	96% identities with *L. anisa*
Growth at 30°C	96% identities with *L. taurinensis*
PHASE II isolate	96% identities with LLAP-9
	96% identities with *L. worsleiensis*

(continued)

Table 4.3 (Continued)

Strain denotation and corresponding water utility	Strain identification (BLAST search)
Y(AZ)/30/1 Water Utility "C" Growth at 30°C PHASE II isolate	99% identities with *L. pneumophila* E5C1 98% identities with *L. pneumophila* subsp. pascullei
Z(AZ)/30/1 Water Utility "C" Growth at 30°C PHASE II isolate	99% identities with *L. anisa* 99% identities with *L. sp.* (strain IBV ATCC 700511) 98% identities with *L. bozemanii* 98% identities with *L. dumoffii*
A(CS5) (Well #5) Water Utility "F" Growth at 30°C PHASE II isolate	98% identities with *L. pneumophila* (several strains)
C(CS5) (Well #5) Water Utility "F" Growth at 30°C PHASE II isolate	98% identities with *L. pneumophila* (several strains)
D(CS5) (Well #5) Water Utility "F" Growth at 30°C PHASE II isolate	98% identities with *L. pneumophila* (several strains)
F(CS5) (Well #5) Water Utility "F" Growth at 30°C PHASE II isolate	97% identities with *L. pneumophila* (several strains)
I(CS5) (Well #5) Water Utility "F" Growth at 30°C PHASE II isolate	98% identities with *L. pneumophila* (several strains)
J(CS5) (Well #5) Water Utility "F" Growth at 30°C PHASE II isolate	98% identities with *L. pneumophila* (several strains)
K(CS5) (Well #5) Water Utility "F" Growth at 30°C PHASE II isolate	98% identities with *L. pneumophila* (several strains)

(continued)

Table 4.3 (Continued)

Strain denotation and corresponding water utility	Strain identification (BLAST search)
A(CS2) (Well #2)	98% identities with *L. pneumophila* (several strains)
Water Utility "F"	
Growth at 30°C	
PHASE II isolate	98% identities with *L. pneumophila* (several strains)
B(CS2) (Well #2)	
Water Utility"F"	
Growth at 30°C	
PHASE II isolate	
LL138-30 (PW_138)	98% identities with *L. parisiensis*
Water Utility "A" (December-January)	98% identities with *L. sp.* (strain IBV ATCC 700511)
Growth at 30°C	98% identities with *L. bozemanii*
PHASE II isolate	98% identities with *L. dumoffii*
M138-30 (PW_138)	98% identities with *L. dumoffii*
Water Utility "A" (January-February)	98% identities with *L. sp.* (strain IBV ATCC 700511)
Growth at 30°C	98% identities with *L. parisiensis*
PHASE II isolate	
WSF	100% identities with "*L. donaldsonii*"
Water Utility "J"	97% identities with *L. londiniensis*
Growth at 30°C	
PHASE II isolate	
HRW	98% identities with *L. dumoffii*
Water Utility "A"	98% identities with *L. gormanii*
Growth at 35°C	98% identities with *L. parisiensis*
PHASE II isolate	
A(PW_103)/35/9	100% identities with *L. erythra* (SE-32A-C8)
Water Utility "A" (February-March)	99% identities with *L. rubrilucens* (WA-270A-C2)
Growth at 35°C	98% identities with *L. taurinensis*
PHASE II isolate	
B(PW_103)/30/9	100% identities with *L. erythra*
Water Utility "A" (February-March)	99% identities with *L. rubrilucens*
Growth at 30°C	98% ientities with *L. taurinensis*
PHASE II isolate	
F(PW_103)/30/9	98% identities with *L. gresilensis*
Water Utility "A" (February-March)	97% identities with *L. beliardensis*
Growth at 30°C	96% identities with *L. micdadei*
PHASE II isolate	96% identities with *L. maceachernii*
F(PW_138)/30/9	99% identities with *L. santicrucis*
Water Utility "A" (February-March)	99% identities with *L. sainthelensi*
Growth at 30°C	99% identities with *L. gratiana*
PHASE II isolate	

(continued)

Table 4.3 (Continued)

Strain denotation and corresponding water utility	Strain identification (BLAST search)
A(PW_138)IMS/30/9	99% identities with *L. dumoffii* (NY-23)
Water Utility "A" (February-March)	98% identities with *L. sp.* (strain IBV ATCC 700511)
Growth at 30°C	98%identities with *L. parisiensis*
PHASE II isolate	
Isolated after IMS	
B(PW_138)IMS/30/9	98% identities with LLAP-2
Water Utility "A" (February-March)	98% identities with LLAP-9
Growth at 30°C	98% identities with *L. rowbothamii* (formerly LLAP-6)
PHASE II isolate	98% identities with *L. lytica* (L2 strain)
Isolated after IMS	97% identities with LLAP-7
G(PW_138)IMS/30/9	99% identities with *L. quateirensis*
Water Utility "A" (February-March)	98% identities with *L. moravica*
Growth at 30°C	97% identities with *L. sp.* (strain IBV ATCC 700511)
PHASE II isolate	97% identities with *L. worsleiensis*
Isolated after IMS	
E(PW_138)IMS/35/9	99% identities with *L.santicrucis*
Water Utility "A" (February-March)	99% identities with *L. sainthelensi*
Growth at 35°C	99% identities with *L. gratiana*
PHASE II isolate	99% identities with *L. sainthelensi*
Isolated after IMS	99% identities with *L. cincinnatiensis*
L(NJ)/35/1	99% identities with *L. sp.* (strain IBV ATCC 700511)
Water Utility "K"	99% identities with *L. anisa*
Growth at 35°C	99% identities with *L. dumoffii*
PHASE II isolate	99% identities with *L. parisiensis*
	99% identities with *L. bozemanii*
A(NJ)IMS/35/1	99% identities with *L. sp.* (strain IBV ATCC 700511)
Water Utility "K"	99% identities with *L. anisa*
Growth at 35°C	99% identities with *L. dumoffii*
PHASE II isolate	99% identities with *L. parisiensis*
Isolated after IMS	99% identities with *L. bozemanii*
DAYS(RI)/30/1	99% identifies with *L. quateirensis*
Water Utility "L"	98% identifies with *L. fallonii* (formerly LLAP –10)
Growth at 30°C	97% identifies with *L. worsleiensis*
PHASE II isolate	97% identifies with *L. dumoffii*
M(RI)/30/1	99% identifies with *L. quateirensis*
Water Utility "L"	98% identifies with *L. fallonii* (formerly LLAP –10)
Growth at 30°C	97% identifies with *L. worsleiensis*
PHASE II isolate	97% identifies with *L. moravica*
I(PW_138)/30/10	97% identities with LLAP-2
Water Utility "A" (April-May)	97% identities with LLAP-9
Growth at 30°C	
PHASE II isolate	

(continued)

Table 4.3 (Continued)

Strain denotation and corresponding water utility	Strain identification (BLAST search)
A(05B)IMS/1	98% identities with *L. fallonii* (formerly LLAP –10)
Water Utility "A"	97% identities with *L. sp.* (strain IBV ATCC 700511)
Growth at 35°C	97% identities with *L. parisiensis*
PHASE II isolate	97% identities with *L. bozemanii*
B(05B)/30/1	97% identities with *L. fallonii* (formerly LLAP –10)
Water Utility "A"	97% identities with *L. sp.* (strain IBV ATCC 700511)
Growth at 30°C	97% identities with *L. parisiensis*
PHASE II isolate	97% identities with *L. bozemanii*
K(05B)/35/1	97% identities with LLAP-10
Water Utility "A"	97% identities with *L. parisiensis*
Growth at 35°C	97% identities with *L. bozemanii*
PHASE II isolate	97% identities with *L. anisa*
A(32G)/30/1	98% identities with *L. anisa*
Water Utility "A"	97% identities with *L. sp.* (strain IBV ATCC 700511)
Growth at 30°C	97% identities with *L. parisiensis*
PHASE II isolate	97% identities with *L. bozemanii*
D(32G)/35/1	98% identities with *L. fallonii* (formerly LLAP –10)
Water Utility "A"	97% identities with *L. sp.* (strain IBV ATCC 700511)
Growth at 35°C	97% identities with *L. parisiensis*
PHASE II isolate	97% identities with *L. bozemanii*
C(94A)/35/1	97% identities with *L. fallonii* (formerly LLAP –10)
Water Utility "A"	97% identities with *L. parisiensis*
Growth at 35°C	97% identities with *L. bozemanii*
PHASE II isolate	97% identities with *L. sp.* (strain IBV ATCC 700511)

Notes: 1. "*L. donaldsonii*" is a not-yet described species as well as *Legionella* sp. (*L. sp.*strain IBV ATCC 700511) but their 16S rRNA gene sequences are available in the database. New species formerly known as *Legionella*-like amoebal pathogens have been recently described (LLAP-1 is the type strain of *L. drozanskii*, LLAP-6 is the type strain of *L. rowbothamii*, LLAP-10 is the type strain of *L. fallonii*) (Adeleke *et al.* 2001). As far as we know, some isolates are still not recognized as separate species (LLAP-2, 7, 9) and are therefore cited by their strain designation.

L. taurinensis is a recently described species closely related to red autofluorescent species (Lo Presti *et al.* 1999). *L. beliardensis* and *L. gresilensis*, two new species isolated from water samples are also recently described (Lo Presti *et al.* 2001).

2. Please note that all the isolates are not indicated in this table. Different strains isolated from the same sample and belonging to the same species are not mentioned in this table.

Additional issues which need to be borne in mind when interpreting the above data are:

- Partial 16S rRNA sequences are not necessarily a sufficient criterion to guarantee species identity (Fox, Wisotzkey, and Jurtshuk, 1992).
- DNA/DNA hybridization or 16S rRNA sequencing of 1,000 or more bases are the optimal methods for determining phylogenetic relationships above the species level (Murray et al. 1990). However, for Stackebrandt and Goebel, only complete sequences (16S rRNA) allow reliable phylogenetic comparison with the available database and caution must be taken when using 16S rRNA gene sequencing as a tool for identification considering that several groups of organisms have been identified sharing almost identical 16S rRNA sequences but in which DNA hybridization is significantly lower than 70% (Stackebrandt and Goebel 1994).
- It is considered that organisms exhibiting less that 97% sequence homology will not reassociate (DNA/DNA hybridization) to more than 60% (Stackebrandt and Goebel 1994). Hence, such isolates cannot be related at the species level. As a consequence, the higher the percentage of homologies for 16S rRNA sequence, the better one can assign a species name to an isolate although even for 16S rRNA identity of 99%, DNA/DNA hybridization may or may not document the existence of species identity (Fox, Wisotzkey, and Jurtshuk, 1992).
- For phylogenetic clustering, please see Appendix F for Phase I isolates and Appendix G for Phase II isolates.

Semi-nested PCR Assay

PCR results (obtained during Phase II) are presented in Tables 4.4 and 4.5 for Water Utility A and Water Utilities C-Q, respectively.

Water Utility A

Results for PCR detection of legionellae are shown in Table 4.4. Lack of detection (both PCR and cultivation) for *Legionella* in several successive samples was rather surprising. It could perhaps have been due to small temperature variations observed during this period, perhaps affecting the presence of *Legionella* in the biofilm growing in the pipe fittings. Or, perhaps it could have been influenced by the use of new pipe materials and a delay in establishing *Legionella* in the biofilm. In order to resolve this issue, two pipe sections were connected in series on each well. One section was removed after one month leaving the remaining pipe section in place for approximately two months. This allowed enough time for *Legionella*, if present, to colonize biofilms or initiate their formation in the pipe. In parallel, whenever possible, the method used during Phase I (removing the end caps and scraping the long-term established biofilm from the well wall) was used in order to see if *Legionella* were still part of the established biofilm. This was conducted on the two selected wells for the first month and the 3 following months. These samples are identified in the table as "end cap" samples versus "pipe" samples.

For PW_103, the presence of PCR inhibitors in the second samples obtained (April-May 2000) was unexpected because no PCR inhibitors were found in the first samples analyzed (March-April 2000). It is possible this could have been due to variations in the local ecosystems (wells) and in water quality. Such variations, including chemical changes, may affect the PCR

Table 4.4
Detection of legionellae by semi-nested PCR (Water Utility A - Florida)

Well ID #	Date Received	Water	Biofilm
103-1*	March-April 2000	-	-
136-1*		-	-
103-2	April-May 2000	Inhibitors	Inhibitors
136-2		Inhibitors	Inhibitors
103-3	May-June 2000	-	-
136-3		N/R	-
103-4	July-August 2000	-	-
138-1		Inhibitors	-
103-5	Aug.-Sept. 2000	-	-
138-2		-	-
103-6	Oct.-Nov. 2000	-	-
138-3		-	-
103-7	Dec.-Jan 2001	+†	+†‡(B and ES)
138-4		+†	+†‡(B and ES)
103-8	Jan.-Feb. 2001	-	Inhibitors‡(B and ES)
138-5		-	Inhibitors‡(B only) ES -
103-9	Feb.-March 2001	-	Inhibitors
138-6		-	Inhibitors
103-10	April-May 2001	-	Inhibitors
138-7		-	Inhibitors
103-11	May-June 2001	N/R	-
138-8		-	-
103-12	June-July 2001	N/R	+§
138-9		-	-
103-13	September 2001	N/R	-
116-1		+§	+§
138-10		+§	+§
116-2	October 2001	+§	+§
138-11		+§	-
KWII RW	November 2000	N/R	+§
KWII IFS	November 2000	N/R	-
05B	April 2001	N/R	-
15A	April 2001	N/R	-
21A	April 2001	N/R	Inhibitors
32G	April 2001	N/R	Inhibitors
72F	April 2001	N/R	Inhibitors
94A	April 2001	N/R	Inhibitors
103 Wiffle Ball	May 2001	N/R	-

* Samples were not enriched before PCR and a conventional DNA extraction was used instead of the QIAGEN DNA extraction kit.

† Samples were positive after second step of the semi-nested PCR procedure.

‡ Two biofilm samples were received for samples 103-7, 103-8, 1384 and 138-5. In each case, B was a PVC pipe section, ES was an end cap scraping resuspended in sterile water.

§ Samples were positive after first step of the semi-nested PCR procedure.

N/R = not recieved.

efficiency (for instance some humic substances are known to inhibit PCR). Whether these variations influenced the presence of *Legionella* is not known.

PW_136 was not investigated anymore following well pumping problems and this well was shut down immediately. A survey was then done on both PW_103 and PW_138, a newly selected well for Phase II. PW_138 was cleaned/rehabilitated during the first week of June 2000 (as mentioned in the well maintenance procedure) and the first pipe fittings for this well were installed the following week. Therefore the first samples investigated were collected one month after this maintenance procedure (July 2000). From the Utility stand point, this is an interesting well to follow because it allowed us to determine how long it takes before a well becomes heavily contaminated with biofilm. Samples obtained since June from this well (maintenance procedure) were found positive for *Legionella* with relatively constant concentrations over time.

Due to technical problems with PW_103, the final stage of Phase II was completed with samples from another well PW_116. PW-103 was investigated over a large period (more than one year from April 2000 to August 2001). IMS – PCR was not undertaken for both PW_138 and PW_116 collected on 12/09/2001 and neither on PW_138 and PW_116 collected on 30/10/2001.

Water Utilities C-Q

Legionella PCR results for water and biofilm samples from utilities C-Q are shown in Table 4.5. In Phase II, the PCR sensitivity was improved by the use of a commercially available DNA extraction kit known to be able to partially remove some PCR inhibitors from different types of samples. The Dneasy extraction kit from Qiagen was used for this purpose.

Overall, the total number of samples positive by PCR is 32/109 (29%) although many samples (21/109) or 19% were found to contain PCR inhibitors. Inhibitors were more common in biofilm samples; 16 out 21 samples containing PCR inhibitors were from biofilms. No attempts were made in this study to remove PCR inhibitors.

Summary for Phase II PCR results:

Number of water samples PCR positive: 15 of 47 (32%)

Number of biofilm samples PCR-positive: 17 of 62 (27%)

Immunomagnetic Separation (IMS) Assays

The IMS – PCR procedure was under development during Phase II of this study in parallel with conventional experiments for *Legionella* detection. Reference strains spiked into natural samples were used for IMS development. In interpreting results for IMS procedures, it should be noted that although attempts to optimize the procedure were partially successful for reference strains, full optimization of an IMS procedure, particularly for field samples, would involve more work and testing on multiple sources that are known to be naturally-contaminated with legionellae. In this study, the IMS procedure was tested on selected Phase II samples in conjunction with both culture and PCR procedures.

IMS is theoretically an ideal concentration tool to selectively remove legionellae (a) from heterotrophic microflora that may overgrow plates during cultivation (IMS cultivation), and (b) from PCR inhibitors which are frequently encountered in environmental samples (IMS-PCR). If successful, IMS could be a useful tool for utilities interested in testing their samples in-house for *Legionella* using a simplified technique that may be able to reduce the complexity of the

Table 4.5
Semi-nested PCR results for water and biofilm samples from water utilities C-Q

Water Utility	Well Identification	Date Received	Water	Biofilm
C (Arizona)	AZ(A)	May 2000	+*	N/R
	AZ(B)		+*	+*
	AZ(BM)		N/R	+*
D (Texas)	EP	May 2000	+*	+*
E (Illinois)	ST	June 2000	+†	-
F (Texas)	CS2	July 2000	+*	+*
	CS5		+*	+*
G (Illinois)	PEO	August 2000	-	-
H (Illinois)	PK	September 2000	+†	+†
I (Missouri)	MEX	October 2000	-	-
J (Michigan)	SF	February 2001	-	-
	MN		-	-
K (New Jersey)	NJ	April 2001	-	-
L (Ontario)	RI	April 2001	-	-
M (Illinois)	CH	May 2001	-	Inhibitors
N (New Brunswick)	NB3	June 2001	+*	+†
	NB7		+†	+†
O (Missouri)	MO18	June 2001	Inhibitors	Inhibitors
	MO34		-	Inhibitors
P (Nebraska)	NE54	July 2001	-	-
	NE66		Inhibitors	-
Q (Florida)	FK8	July 2001	-	N/R
	FK13		-	N/R

* Samples tested positive during the first step of the semi-nested procedure.

† Samples tested positive during the second step of the semi-nested procedure.

N/R = not received.

decontamination and culture methods, and or reduce or remove PCR inhibitors. Results for IMS-cultivation for legionellae are presented in Tables 4.6 and 4.7 for Water Utility A and for Water Utilities K-Q, respectively. Results for IMS-PCR for legionellae are presented in Tables 4.8 and 4.9 for Water Utility A and for Water Utilities K-Q, respectively.

As a result of the ongoing method development, the preparation of the water and biofilm samples for IMS, as well as the IMS procedure, were modified during the study. IMS carried out between August and November 2000 yielded no positive colonies; these are not reported below. All the plates were overgrown with contaminating bacteria making it impossible to visually identify *Legionella* colonies if present. The IMS procedure was altered for samples received in January and February 2001. Storing the samples at 4°C overnight instead of using the enrichment combined with heat and acid treatments did reduce the amount of overgrowth on the plates. Even so, no positive colonies were isolated, although positive colonies were found from some samples

Table 4.6
IMS-cultivation on selected samples (Phase II - water utility "A")

Well ID	Date	Water (CFU/L)	Strain identification	Biofilm (CFU/cm^2)	Strain identification
PW_103-9	Feb-Mar 2001	None	N/A	None	N/A
PW_138-9	July 2001	40	*L. dumoffii*	None	N/A
		40	LLAP-2		
		40	*L. santicrucis*		
		40	*L. quateirensis*		
PW_103-10	Apr-May 2001	None	N/A	None	N/A
PW_138-10	Sep 2001	None	N/A	None	N/A
PW_103-11	May-Jun 2001	N/R	N/A	5.2 (C, UC)	*L. quateirensis*
				1.3 (C)	*L. erythra*
				5.2 (UC)	*L. dumoffii*
PW_138-11	October 2001	None	N/A	None	N/A
PW_103-12	July 2001	N/R	N/A	5.8	*L. quateirensis*
PW_138-12	Nov 2001	2.5×10^2	*L. quateirensis*	None	N/A
05B	Apr 2001	N/R	N/A	1.3×10^3 CFU/80 mL	LLAP-10
15A	Apr 2001	N/R	N/A	None	N/A
21A	Apr 2001	N/R	N/A	None	N/A
32G	Apr 2001	N/R	N/A	None	N/A
72F	Apr 2001	N/R	N/A	None	N/A
94A	Apr 2001	N/R	N/A	None	N/A

C = Combined IMS, UC = Uncombined IMS, N/R = not received, N/A = not available.

using conventional cultivation. The final significant change to the procedure involved the reduction in the concentration of the primary monoclonal antibody to 1 µg/mL. With these final alterations, *Legionella* colonies were isolated from both water and biofilm samples as indicated in Tables 4.6 and 4.7. Uncombined IMS was employed for samples received up until May 2001. Since tests using reference strains of *Legionella* showed that combined IMS was as effective as uncombined IMS with regards to recovery of samples (data not shown), environmental samples received in May underwent both types of IMS. Similar results were found using both methods. Samples received in June and July 2001 were subjected only to the combined IMS procedure.

Using IMS-cultivation, 3/16 water samples (19%), and 5/23 biofilm samples (22%) contained detectable legionellae. Overall, 21% (8/39) of samples were found positive for *Legionella*. In view of the preliminary nature of this study this was considered quite encouraging and further method development for *Legionella* IMS may be valuable. Even if the antibodies used may not turn out to be highly specific for legionellae, they may still significantly reduce the interfering bacteria and facilitate culture. In the IMS-PCR, 9/32 water samples (28%), and 6/45 biofilm samples (13%) were positive for legionellae. Overall the percentage of positive samples was 19% (15/77). The apparently more successful application of IMS-PCR to the water samples may reflect a better ability of the IMS to be effective in the absence of large quantities of particulate materials.

For the most part, IMS – PCR was found negative on both PW_103 and PW_138 (Water Utility A) water and biofilm samples tested. For the April-May 2001 samples, IMS-PCR was

Table 4.7
IMS-cultivation on selected samples (Phase II - utilities K-Q)

Water utility	Well ID	Water (CFU/L)	Strain identification	Biofilm (CFU/cm^2) or as stated	Strain identification
K (New Jersey)	NJ	None	N/A	5.7	*L.* sp. ATCC 700511
L (Ontario)	RI	None	N/A	None	N/A
M (Illinois)	CH	None	N/A	None	N/A
N (New Brunswick)	NB#3	None	N/A	None	N/A
	NB#7	None	N/A	None	N/A
O (Missouri)	MO#18	None	N/A	4.1 x 10^3 and 2.0 x 10^3 CFU in 250 mL	LLAP *L. quateirensis*
	MO#34	N/A	N/A	None	N/A
P	NE#54	None	N/A	None	N/A
(Nebraska)	NE#66	None	N/A	None	N/A
Q	FK#8	None	N/A	N/A	N/A
(Florida)	FK#13	25	*L. pneumophila*	N/A	N/A

N/R = not received, N/A = not available.

found positive for the PW_103 water sample and the PW_138 water and biofilm samples. On this one occasion there appeared to be a significant advantage of IMS-PCR because the normal PCR did not detect legionellae in water and both biofilm samples had PCR inhibitors. However, the IMS – PCR procedure was not able to overcome the problem of PCR inhibitors during the IMS-PCR. When comparisons with the semi-nested PCR alone are made, it is clear that some samples which showed inhibitors in PCR do not have them in IMS – PCR. Conversely, some negative samples in the simple semi-nested procedure that showed no PCR inhibition, showed inhibition in the IMS – PCR reaction. The latter result is particularly unexpected and does not have an immediate explanation. Clearly the samples are very complex natural matrices and there are a lot of unknowns and potential interferences that may preclude molecular reactions, including those that depend on reactions with antibodies, from being relied on routinely.

Problems using IMS-cultivation for recovery of environmental isolates appear to be common. In an unrelated experiment using *Cryptosporidium parvum*, IMS with spiked environmental samples gave efficient recovery of microorganisms although naturally occurring organisms were not recovered from unspiked samples in all investigations (Rochelle et al. 1999). Other organisms have been successfully isolated from samples using IMS-cultivation (Uyttendaele, Van Hoorde, and Debevere, 2000). To the research team's knowledge, neither IMS-cultivation nor IMS – PCR using *Legionella* as the target organism have been attempted previously. The results above, using the final alterations to the IMS procedure, show that it is possible to use IMS to isolate *Legionella* spp. from environmental samples. In this study, 18% of the samples were found to be positive using IMS-cultivation as a detection procedure. The IMS-PCR procedure was less successful overall but could perhaps be improved with further optimization and inclusion of specific inhibitor removal steps.

The design of the IMS procedure lowers the detection limit by 10 fold over the conventional cultivation techniques. This increase in sensitivity could allow for detection of

Table 4.8
IMS-PCR results for *Legionella* (Water Utility A)

Well identification	Date received	Water	Biofilm
103-1	March-April 2000	ND	ND
136-1		ND	ND
103-2	April-May	ND	ND
136-2	2000	ND	ND
103-3	May-June	ND	ND
136-3	2000	ND	ND
103-4	June-July 2000	-	-
138-1		-	-
103-5	Aug.-Sept. 2000	-	-
138-2		-	-
103-6	Nov.-Dec. 2000	-	-
138-3		-	-
103-7	Dec.-Jan.2000	-	- (B and ES)*
138-4		-	- (B and ES)*
103-8	Jan.-Feb. 2001	-	- (B and ES)*
138-5		-	- (B and ES)*
103-9	Feb.-March 2001	-	Inhibitors
138-6		-	Inhibitors
103-10	April-May 2001	+†	Inhibitors
138-7		+†	+†
103-11	May-June 2001	N/R	- (UC and C)
138-8		-	-(UC), inhib(C)
103-12	June-July 2001	N/R	-
138-9		-	Inhibitors
116-1	September 2001	ND	ND
138-10		ND	ND
138-11	October 2001	ND	ND
116-2		ND	ND
KII RW	April 2001	N/R	-
KII IFS	April 2001	N/R	-
05B	April 2001	N/R	-
15A	April 2001	N/R	-
21A	April 2001	N/R	-
32G	April 2001	N/R	Inhibitors
72F	April 2001	N/R	Inhibitors
94A	April 2001	N/R	-
103 Wiffle	April 2001	N/R	-

* (B and ES) indicate 2 separate biofilm samples received, B designates the PVC pipe section and ES designates biofilm received from scraping the end-cap (as was done in Phase I).

† Samples were positive during the first step of the PCR procedure.
C = Combined IMS; UC = Uncombined IMS; ND = not done; N/R = not received.

Table 4.9
IMS-PCR results for *Legionella* (water utilities C to Q)

Water utility	Well ID	Date	Water	Biofilm
C (Arizona)	-	May 2000	ND	ND
D (Texas)	-	May 2000	ND	ND
E (Illinois)	-	June 2000	ND	ND
F (Texas)	CS2	July 2000	+*	+*
	CS5		+*	+*
G (Illinois)	PEO	Aug 2000	-	Inhibitors
H (Illinois)	PK	Sep 2000	+†	+†
I (Missouri)	MEX	Oct 2000	-	-
J (Michigan)	SF	Feb 2001	ND	ND
	MN		ND	ND
K (New Jersey)	NJ	Apr 2001	-	Inhibitors
L (Ontario)	RI	Apr 2001	+	-
M (Illinois)	CH	May 2001	-(C), +(UC)	-
N (New Brunswick)	NB3	Jun 2001	+	+
	NB7		+	+
O (Missouri)	MO18	Jun 2001	Inhibitors	Inhibitors
	MO34		-	-
P (Nebraska)	NE54	Jul 2001	-	-
	NE66		-	-
Q (Florida)	FK8	Jul 2001	-	N/R
	FK13		Inhibitors	N/R

* Samples positive during the first step of the PCR procedure.

† Samples positive during the second step of the PCR procedure.

C = Combined IMS; UC = Uncombined IMS; N/R = not received; ND = not done.

bacteria that are on the borderline of detection using conventional cultivation. Experiments using *Listeria monocytogenes* showed that although the detection limit was lowered, the ratio of *L. monocytogenes* to non-*Listeria* flora was not increased (Uyttendaele, Van Hoorde, and Debevere, 2000). These findings are similar to ours using *Legionella* since the amount of overgrown plates using the IMS procedure was 22.2% compared to 15.9% found using conventional cultivation. Whether or not these speak to a lack of specificity of the antibody remains to be determined.

The potential advantages offered by the addition of IMS to either culture or PCR detection of legionellae are significant, and worthy of more thorough investigation, especially if some truly *Legionella* specific antibodies can be obtained.

Summary of *Legionella* Data and Method Comparisons

To simplify some comparisons of the data using different detection methods, all the results for *Legionella* detection have been combined into a single summary table (Table 4.10). The wells comprise a range of different well water temperatures with *Legionella* being found in all temperature ranges tested. Well depths are included in this table just to show that legionellae are recovered even from wells that are more than 1000 feet deep.

Comparison of IMS and Conventional Cultivation and PCR Techniques

The purpose of using IMS during this study was to determine if using IMS isolation of bacterial cells prior to cultivation or PCR enhanced detection of *Legionella* in environmental water and biofilm samples. As seen in the conventional cultivation and semi-nested PCR results, neither of these methods separately detected *Legionella* in all *Legionella*-positive samples. By combining these techniques with IMS *Legionella* was detected in a higher number of samples. When all methods tried on any sample are combined, 63% of water samples and 44% of biofilm samples were positive for legionellae.

IMS-Cultivation Versus Conventional Cultivation

When comparing the IMS-cultivation to the conventional cultivation technique several observations were made. Using IMS-cultivation, *Legionella* were isolated from only 19% of water samples compared with 48% using conventional cultivation. Percentages for biofilm samples were quite similar whether conventional cultivation was used (23%) or IMS-cultivation (22%). The major problem with conventional cultivation is overgrowth by heterotrophic bacteria. In theory, IMS should reduce contaminating bacteria and, therefore, the numbers of overgrown plates. The IMS-cultivation results in this study do not support this theory; 22.2% of IMS-cultivation plates were overgrown compared to 15.9% for conventional cultivation. It should be noted (Table 4.10) that different numbers of samples were used and different numbers of plates were inoculated for each technique. Conventional cultivation used direct plating, heat treatment and acid treatment on the three media (36 plates for each sample). IMS-cultivation only used direct plating and acid treatment (12 plates for each sample). It is also possible that the monoclonal antibody was unable to bind to certain types of environmental *Legionella,* or the surface epitopes of the bacteria were not in a condition to bind the Mab. The percentages of samples positive by cultivation and IMS-cultivation were quite similar, regardless of whether the full sample set was considered or just a selected subset after the time that the IMS procedure was optimized. Although not conclusive, this suggests that optimization steps successful on reference strains in the laboratory offered no advantage when dealing with field samples.

IMS-PCR Versus Conventional Semi-nested PCR

If the results of *Legionella* analysis using semi-nested PCR with enrichment but no prior IMS are compared with those of IMS-PCR the following was observed. The percentages of positive samples are quite similar for water although the numbers tested are quite different. For biofilm samples, the percentage positive for legionellae is lower for the IMS-PCR when the whole data set is considered. However, the results are much more similar when based on the restricted data set (Table 4.10). Although the IMS culture and the IMS PCR added additional positive samples to those that were positive for legionellae, they were few in number compared to those identified using the conventional techniques without IMS.

Table 4.10
Summary of Phase II *Legionella* results by utility for all methods used

			Date	Culture (temp°C)		IMS-Culture		PCR		IMS-PCR	
Utility	Well-sample ID	Depth (ft)	2000-2001	Water/L	Biofilm	Water	Biofilm	Water	Biofilm	Water	Biofilm
A	103-1	767	Mar/Apr/00	280(30)	0	nd	nd	-	-	nd	nd
A	103-2		Apr/May/00	0	1/cm^2	nd	nd	I	I	nd	nd
A	103-3		May/Jun/00	0	0	nd	nd	-	-	nd	nd
A	103-4		Jul/Aug/00	0	0	nd	nd	-	-	-	-
A	103-5		Aug/Sep/00	0	0	nd	nd	-	-	-	-
A	103-6		Oct/Nov/00	0	0	nd	nd	-	-	-	-
A	103-7		Dec/Jan/01	0	0	nd	nd	+	+/+	-	-/-
A	103-8		Jan/Feb/01	0	0	nd	nd	-	I/I	-	-/-
A	103-9		Feb/Mar/01	250(30) 250(35)	0	0	0	-	I	-	I
A	103-10		Apr/May/01	100(35)	0	0	0	-	I	+	I
A	103-11		May/Jun/01	NR	35/cm^2	NR	12/cm^2	NR	-	NR	-
A	103-12		Jul/01	NR	44/cm^2	NR	6/cm^2	NR	+	NR	-
A	103-13		August/01	NR	0	NR	nd	NR	nd	NR	nd
A	136-1	780	Mar/Apr/00	0	0	nd	nd	-	-	nd	nd
A	136-2		Apr/May/00	0	0	nd	nd	I	I	nd	nd
A	136-3		May/Jun/00	0	0	nd	nd	nd	-	nd	nd
A	138-1	778	Jul/Aug/00	431(30)	0	nd	nd	I	-	-	-
A	138-2		Aug/Sep/00	413(30)	0	nd	nd	-	-	-	-
A	138-3		Oct/Nov/00	0	0	nd	nd	-	-	-	-
A	138-4		Dec/Jan/01	4600(30)	0	nd	nd	+	+/+	-	-/-
A	138-5		Jan/Feb/01	206(30)	0	nd	nd	-	-/I	-	-/-
A	138-6		Feb/Mar/01	200(30) 306(35)	0	nd	nd	-	I	-	I
A	138-7		Apr/May/01	413(30)	0	nd	nd	-	I	+	+
A	138-8		May/Jun/01	0	0	nd	nd	-	-	-	-

(continued)

Table 4.10 (Continued)

Utility	Well-sample ID	Depth (ft)	Date 2000-2001	Culture (temp°C)		IMS-Culture		PCR		IMS-PCR	
				Water/L	Biofilm	Water	Biofilm	Water	Biofilm	Water	Biofilm
A	138-9		Jul/01	619(35)	0	160	0	-	-	-	I
A	138-10		Sep/01	100(35)	0	0	0	+	+	nd	nd
A	138-11		Oct/01	25(30) 100(35)	0	0	0	+	-	nd	nd
A	138-12		Nov/01	nd	nd	250	0	nd	nd	nd	nd
A	116-1		Sep/01	100(35)	0	nd	nd	+	+	nd	nd
A	116-2		Oct/01	50(35)	0	nd	nd	+	+	nd	nd
A	KWII-RW		Nov/00	NR	206(30) 100(35)	NR	nd	NR	+	NR	-
A	KWII-IFS		Nov/00	NR	0	NR	nd	NR	-	NR	-
A	05B		Apr/01	NR	206(30) 500(35)	NR	1300/80 mL	NR	-	NR	-
A	15A		Apr/01	NR	0	NR	0	NR	-	NR	-
A	21A		Apr/01	NR	0	NR	0	NR	I	NR	-
A	32G		Apr/01	NR	206(30) 200(35)	NR	0	NR	I	NR	I
A	72F		Apr/01	NR	0	NR	0	NR	I	NR	I
A	94A		Apr/01	NR	100(35)	NR	0	NR	I	NR	-
A	103 whiffle ball		May/01	NR	nd	NR	nd	NR	-	NR	-
C	AzA	1100	May/00	100(30)	NR	nd	NR	+	NR	nd	NR
C	AzB	1100	May/00	200(30)	0	nd	nd	+	+	nd	nd
C	AzBM	1100	May/00	NR	13(30) 25(35)	NR	nd	NR	+	nd	nd
D	EP	150-1000	May/00	0	0	nd	nd	+	+	nd	nd
E	ST	1800	Jun/00	0	0	nd	nd	+	-	nd	nd

(continued)

Table 4.10 (Continued)

Utility	Well-sample ID	Depth (ft)	Date 2000-2001	Culture (temp°C) Water/L	Culture (temp°C) Biofilm	IMS-Culture Water	IMS-Culture Biofilm	PCR Water	PCR Biofilm	IMS-PCR Water	IMS-PCR Biofilm
F	CS2	2975	Jul/00	100(30)	10(30)	nd	nd	+	+	+	+
		2975		100(35)	5(35)						
F	CS5	2975	Jul/00	183(30)	5(30)	nd	nd	+	+	+	+
G	PEO	162	Aug/00	0	0	nd	nd	-	-	-	I
H	PK	85	Sep/00	0	0	nd	nd	+	+	+	+
I	MEX	na	Oct/00	0	0	nd	nd	-	-	-	-
J	SF	92	Feb/01	100(30)	0	nd	nd	-	-	nd	nd
J	MN	125	Apr/01	0	0	nd	nd	-	-	nd	nd
K	NJ	598	Apr/00	0	118 (35)	0	6	-	-	-	I
L	RI	~200	Apr/00	412(30)	6200/L (30)	0	0	-	-	+	-
M	CH	250-300	May/01	0	0	0	0	-	I	+	-
N	NB3	107	Jun/01	0	0	0	0	+	+	+	+
N	NB7	117	Jun/01	0	0	0	0	+	+	+	+
O	MO18	125	Jun/01	300(30)	0	0	6000 per 250 mL	I	I	I	I
				100(35)							
O	MO34	76	Jun/01	0	0	nd	0	-	I	-	-
P	NE54	76	Jul/01	0	0	0	0	-	-	-	-
P	NE66	69	Jul/01	512(30)	14(30)	0	0	I	-	-	-
Q	FK8	50	Jul/01	0	NR	0	NR	-	NR	-	NR
Q	FK13	50	Jul/01	800(35)	NR	25	NR	-	NR	I	NR

(continued)

Table 4.10 (Continued)

Utility	Well-sample ID	Depth (ft)	Date 2000-2001	Culture (temp°C)		IMS-Culture		PCR		IMS-PCR	
				Water/L	Biofilm	Water	Biofilm	Water	Biofilm	Water	Biofilm
Using all data											
Number samples positive				23	13	3	5	15	17	9	6
Number samples tested				48	57	16	23	47	62	32	45
% positive				48	23	19	22	32	27	28	13
Using only those samples where IMS was used after final changes to IMS procedure [A138-8 to A138-11 & Wells M-Q]											
Number samples positive				6	1	2	1	4	3	3	2
Number samples tested				13	11	11	10	13	11	11	9
% positive				46	9	18	10	31	27	27	22
					WATER				**BIOFILM**		
***Samples positive by any method tried**					**30**				**24**		
***Samples negative by all methods tried**					**18**				**33**		
% positive samples					**63%**				**42%**		

Note: For PCR and IMS-PCR, when two results are shown in the same cell – two samples were processed. See Tables 4-4 and 4-8.

* Based on Sample ID only, duplicates from one ID# not included.

NR = not received; na = not available; nd = not done; I = inhibitors present.

As discussed earlier, the numbers of samples having PCR inhibitors was similar for both PCR procedures, although inhibition in the two techniques was not necessarily seen in the same samples. Improvements are necessary for both procedures to try and eliminate these from the samples. For IMS, an increased number of washing steps could reduce inhibitors. Post IMS treatments to reduce inhibitors could be applied to both samples. These may include changes to the DNA extraction procedure or additional clean-up steps.

All Methods Combined

Table 4.10 shows that if all methods are combined legionellae were present in 63% of water samples and 42% of biofilms. Clearly the numbers of legionellae detected are at a minimum of those present. This can be attributed to three principal reasons: a) overgrown plates made colony selection difficult, and impossible in some samples, b) observed variation in samples from utility A wells suggests that wells negative on a single sample may show legionellae if subsequent samples were to be taken, and c) the significant numbers of samples containing PCR inhibitors. If just wells are considered, with multiple samples from a single well being combined, then only 7 of 27 wells from 16 utilities were negative for legionellae. Two utilities had negative samples only: both submitted a single sample from one well. Five utilities submitted samples from more than one well where at least one well was positive. The majority of utilities (9) submitted one or more samples that all contained legionellae.

PROTOZOA ANALYSIS (ON SELECTED SAMPLES)

Results have been summarized in Table 4.11. The numbers of samples examined for protozoa was relatively small compared to the total numbers tested. The results in Table 4.11 are expanded by additional pertinent observations for each utility tested in Appendix H. To our knowledge, *Colpoda* have never been found (described) to promote the growth/survival of *Legionella*. Therefore it may be interesting to try to investigate whether this ciliate can act as a natural amplifier in nature and in groundwater in particular.

Direct microscopic observation of all four samples from water Utility F (two filtered water samples and two biofilms) before and after centrifugation, at both high and low speeds, showed an interesting organism in the two biofilm samples. They were green, indicating chlorophyll, 3 μm in size and had a small knob at one end - no flagella were seen arising from the knob even at 1000x magnification. They were not like any protozoa that have been historically associated with *Legionella* (*Naegleria, Acanthamoeba* and *Hartmannella* are all much larger and don't have chlorophyll). These cells appeared to die off in culture media specifically designed for growing free living protozoa after 72 hours. No protozoa that were historically described to be associated with *Legionella* were found in Utility F samples. The ecosystem where these samples were obtained is warm (40 and 45°C). Free-living protozoa seldom survive at 36-40°C though some Japanese workers in the 1930s found that some protozoa could be acclimatized to temperatures up to 50°C.

Legionella survival in water environments can be highly dependent, as previously described, on algae (Tison, Baross, and Seidler, 1983; Hume and Hann 1984; Tison 1987) as well as cyanobacteria (Berendt 1981; Bohach and Snyder 1983a; Bohach and Snyder 1983b). Legionellae were found by all methods used for Utility F samples, and *L. pneumophila* were confrirmed in the culture samples. It is not known if the green algae observed may support the intracellular multiplication of *Legionella*, or if the high temperatures in Utility F wells permitted *Legionella* growth to be sustained, even in the absence of suitable host organisms.

Table 4.11
Results of protozoa analysis for selected water utilities

Water utility	Sample ID	Date received	*Legionella* analysis		Protozoa analysis	
			Water	Biofilm	Water	Biofilm
A	103-4	June-July 2000	None	None	Algae	Algae
	138-1		Present	None	Ciliates, trophozoites	Algae
	103-5	Aug.-Sept. 2000	None	None	*Colpoda*	None
	138-2		Present	None	None	None
	103-6	Nov.-Dec. 2000	None	None	None	None
	138-3		None	None	None	None
	103-7	Dec.-Jan. 2000	None	None	None	Flagellates
	138-4		Present	None	None	*Hartmannella*, trophozoites
	103-8	Jan.-Feb. 2001	None	None	None	None
	138-5		Present	None	None	None
	103-9	Feb.-March 2001	Present	None	Flagellates	*Naegleria*, trophozoites, flagellates
	138-6		Present	None	*Colpoda*, ciliates	*Naegleria, Colpoda,* amoebic cysts, trophozoites
	KII RW	April 2001	N/A	Present	N/A	Ciliates (probably *Litonitus* or *Acineria**; *Chaetopsira, Uroleptus*), phytoflagellates (probably *Chromulina, Chilomonas,Cryptomonas*), cysts
	KII IFS	April 2001	N/A	None	N/A	None
F	CS2	July 2000	Present	Present	None	Algae
	CS5		Present	Present	None	Algae
H	PK	September 2000	None	None	None	None
I	MEX	October 2000	None	None	None	None
J	MN	February 2001	None	None	None	None
	SF		Present	None	None	None
K	NJ	April 2001	None	Present	None	None
L	RI	April 2001	Present	Present	None	None

* *Acineria* mentioned in Foissner (1996). *Litonitus* and *Acineria* are very difficult to differentiate.

Overall, the presence of protozoa that are known to support *Legionella* growth and that may promote *Legionella* survival in surface waters was found marginally in the samples investigated. During the survey conducted on water utility "A" samples, *Hartmannella* and possibly *Tetrahymena* were observed. Numerous other protozoa including unidentified ciliates and flagellates and amoebal cysts were seen. Some protozoa (*Microthorax*, *Colpoda*) that have not been yet associated with the presence of *Legionella* in aquatic habitats were also found. Additional testing would be beneficial to identify if these protozoa may be considered as possible hosts for *Legionella* in aquatic habitats. The number of protozoan species found to support *Legionella* growth is constantly increasing (Table 1.2) despite a general lack of interest in *Legionella* ecology. Investigating the presence of protozoa in water systems is important, since they are likely to represent the main way of colonizing and recolonizing water systems with legionellae.

Green algae have also been reported in a number of samples, though it is not known if they may support the intracellular multiplication of legionellae in groundwater.

Although all samples received were not investigated for the presence of protozoa, it is very likely that some groundwater systems may be colonized by protozoa known to support *Legionella* growth. These protozoa may be responsible for the spread and survival of *Legionella* in water distribution systems, probably through the chlorine-resistant cysts that may harbor and protect *Legionella*.

PARAMETERS INFLUENCING THE PRESENCE OR ABSENCE OF *LEGIONELLA*

In the absence of adequate and simple bacteriological indicators of legionellae, there is an incentive to evaluate other chemical, hydrogeological, geographic or demographic parameters that may be used to assess the vulnerability of wells to legionellae or other contaminants potentially originating from surface soils or waters. The questionnaire "Well Information Checklist" and sample information "Sample and Data Sheet" sent to participating utilities (Appendix B and C, respectively) asked a number of pertinent questions in this regard. However, perhaps because of the length of the questionnaire, and or the difficulties in answering some of the questions by personnel who performed the sample collection, relatively little data was completed and hence available for analyses. The main data sets that could be investigated for relationships to *Legionella* presence, and possible use as indicator(s) of *Legionella* colonization in groundwater systems, were for standard parameters or chemical analyses performed by the participating utilities. Even these data sets are incomplete for some utilities. Thus the analysis which could be done was severely restricted in potential predictive value.

However, it should also be borne in mind that the widespread occurrence of legionellae in the majority of wells and most of the utilities reported above make it less likely that a single indicator can be found which would be sufficiently predictive of *Legionella* presence. This is also supported by the isolation of legionellae from groundwater wells exhibiting very different water temperature and pH values as indicated in the Figure 4.1.

Data Selection

From a biostatistical point of view, primary observation of the data generated through this study led to the definition of two strategies for analysis:

A first data set was selected and submitted for analysis. This included a table of 14 parameters for which 30 observations (measurements) were done. Parameters included in this first

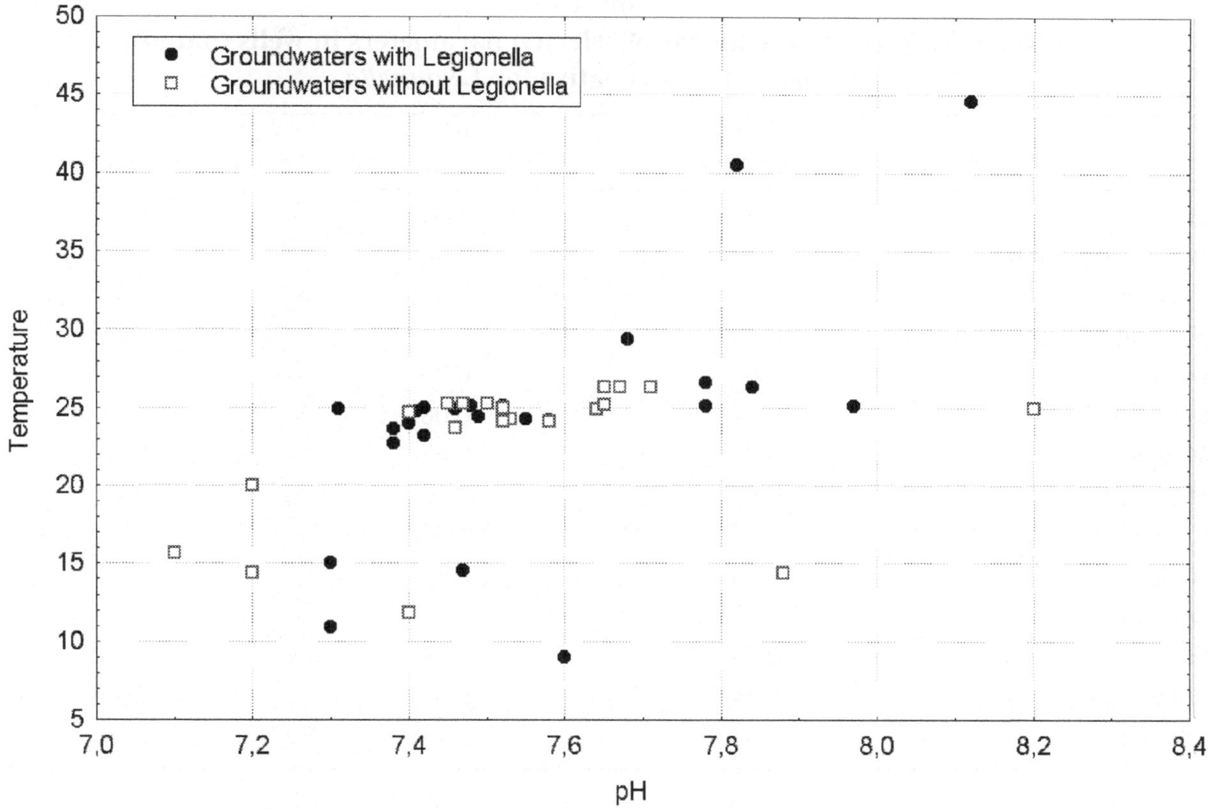

Temperature is indicated in degrees Celsius, pH was measured as standard pH units

Figure 4.1 Influence of water temperature and pH on *Legionella* distribution in selected groundwater wells

batch were: water temperature, pH, conductivity, turbidity, dissolved oxygen, total organic carbon, and iron, manganese, magnesium, copper, nickel, sodium, sulfate and calcium concentrations.

A second set of data was then selected and submitted for analysis. This second set was made of 5 parameters (water temperature, pH, conductivity, turbidity and dissolved oxygen) for which 44 observations (measurements) were obtained from participating water utilities.

Analyses and Discussion

On both sets of data, several analytical methods were selected to find potential correlations between some of the parameters investigated and the presence of *Legionella* in the corresponding well.

First, simple variance testing and discriminative factorial analysis were used for analysis. For both methods of analysis, no discrimination was obtained between positive and negative wells for *Legionella*. From a biostatistical point of view, this may be regarded as a non-significant analysis, due to a poor representation of the parameters, meaning that the number of observations available for the 14 parameters selected in the first set of data generated is not sufficient. Overall, the conclusion is the same for the second set of data selected despite the presence of an additional

Table 4.12
Differences between the means of selected parameters in wells that are positive versus negative for *Legionella*

	MP (*Legionella*+)	MA (*Legionella*-)	*p* values
pH	7.55	7.56	90%
Water temperature	4.73	4.47	50%
Conductivity	3.41	3.52	78%
Turbidity	0.75	0.84	80%
Dissolved oxygen	0.67	0.79	74%
Iron	0.57	0.08	19%
Manganese	0.24	0.62	30%
Magnesium	0.69	0.23	22%
Copper	0.48	1.01	15%
Nickel	0.36	0.92	13%
Sodium	0.57	0.31	49%
Sulfate	0.38	0.54	66%
Total organic carbon	6.34	6.38	93%
Calcium	2.02	2.13	78%

44 observations. Schematic representations of factorial analyses for these two sets of data generated by the scatters module from ADE 4.0 are provided in Appendix I. Therefore, for the parameters investigated and using simple variance testing and discriminative factorial analysis, no indicator of the presence of *Legionella* was identified.

Data were also investigated using linear regression methods. ADE 4.0 provides several linear regression methods. These methods are univariate regression, orthogonal regression and linear regression. All data collected cannot differentiate between positive and negative wells for *Legionella* when considering logistical regression analyses. Also, a logistical regression analysis combining 5 parameters (pH, temperature, conductivity, turbidity and dissolved oxygen) fails to accurately predict whether a well will be positive or negative for *Legionella* in 93.8% of cases. Therefore, these parameters cannot act as indicators of the presence of *Legionella* within the wells investigated and using the statistical analysis methods selected (logistical regression analysis using the Quasi-Newton parameter estimation method).

In addition, a Student t-test analysis was also conducted to compare the influence of several parameters taken individually on the presence of *Legionella* in groundwater. In the following list, the means for selected parameters (pH, water temperature, conductivity, turbidity, dissolved dxygen, iron, manganese, magnesium, copper, nickel, sodium, sulfate and total organic carbon concentrations) have been calculated for *Legionella*-positive wells positive wells (designated as MP) and *Legionella*-negative wells (designated as MA). When testing each parameter individually with the Student t-test, no significant differences between positive and negative wells were observed (see *p* values in Table 4.12).

A difference is considered significant when p values are < 5%. Although there are no statistically significant associations identified in Table 4.12, some weak positive associations may be predicted for iron and magnesium, and some weak negative associations for manganese, copper and nickel. These differences may deserve further investigations but it should be done only in the context of *Legionella* ecology in the presence of both protozoan hosts and competing natural bacterial populations.

Effects of positive correlations would be likely to enhance growth of legionellae and/or their protozoan hosts. Conversely, negative correlations would be likely to reduce growth of either legionellae and/or their natural protozoan hosts. That other microorganisms which are potential predators or competitors for legionellae may also be affected, though not necessarily similarly, suggests a complex ecological picture. Any studies investigating this issue in more depth should therefore be conducted under as natural conditions as possible.

The issue of iron and copper as potential positive and negative influences, respectively, is of particular interest to the drinking water industry because of the use of iron in municipal infrastructure and copper in domestic supplies. Studies to explore these influences should therefore have direct relevance to *Legionella* occurrence in municipal water supplies whether they originate from surface of groundwater sources.

The iron acquisition function of legionellae is well recognized to play an important role in pathogenesis (Cianciotto, 2001), but the role that iron may play in the ecology of legionellae or their persistence in the environment as aquatic organisms is not clear. In fact, a survey of Nova Scotia hospitals (Marrie et al. 1994) for legionellae concluded that Legionella-positive hospitals were older and larger with total system recirculation. Water from these hospitals also had a higher pH and calcium content but lower sodium, potassium, nitrate, iron and copper. Thus high concentrations of iron per se may not be essential for legionellae to be present. Another study on materials in water supply systems (Pongratz et al. 1994) supports the iron and copper linkages identified above. However, iron was not more prone to be colonized by legionellae than polyethylene, but copper was significantly less likely to be colonized. It is possible therefore that copper plumbing at the user interface to the central water system may offer some protection to the end water user. The use of copper as part of copper-silver ionization systems (Lin et al. 1996; 1998) to control *Legionella* in institutions further supports its antagonistic role towards legionellae and or protozoa, although the efficacy of such a system may be compromised depending on water pH (Lin et al. 2002).

DISCUSSION AND CONCLUDING REMARKS FOR PHASE II

At the start of Phase II of this study, the objectives included determining how widespread was *Legionella* contamination of groundwater. Table 4.10 shows that if all methods are combined legionellae were present in 63% of water samples and 42% of biofilms. Clearly the numbers of legionellae detected are at a minimum of those present. This can be attributed to four reasons: a) the overgrown plates made colony selection difficult and impossible in some samples, b) the variation observed in samples from wells of utility A suggest that some of the wells which were negative on a single sample may show legionellae if subsequent samples were to be taken, c) the significant numbers of samples containing PCR inhibitors, and d) the general difficulties in isolating legionellae from biofilm samples with heavy precipitates and particulates – see description of samples in Appendix D.

Another objective was to determine the usefulness of applying an additional separation/concentration method, in this case IMS, to improve the cultivation and PCR methods currently used. The results of the study so far indicate that the IMS could not produce in the field samples any increase in the percentage of positive samples. Thus, unless significant improvements can be made to the IMS procedure, and it can be properly validated with field samples known to be positive for a wide variety of *Legionella* species, it may not be worth including this extra step.

Although it was anticipated that Legionella would be found even in the absence of fecal indicators, the data in this study support this conclusion.

Since *Legionella* has previously been shown to be ubiquitous in surface water samples, it was expected that this bacterium would also be present in groundwaters. Its presence in both warm and cold groundwaters was established in this study and so temperature was not a useful indicator of its presence. Cultivation was chosen as one of the primary detection methods since it is currently the "gold standard" method for detection of *Legionella*. A semi-nested PCR approach was also used since *Legionella* are difficult to cultivate and are thought to be predominately in a viable but not culturable state. PCR is capable of detecting both the culturable as well as the non-culturable cells. As seen in previous papers, when comparing cultivation to PCR, the amount of samples positive by cultivation is usually much lower (Lye et al. 1997). In this study, the number of positive samples by conventional cultivation and the number of positive samples by semi-nested PCR were approximately equal, though with higher numbers culture positive in water and higher numbers PCR positive in biofilms. One possible explanation of this finding is that although *Legionella* were present in the groundwater samples, they were in low concentrations. Although it is difficult to 'pick' legionellae from among other contaminants, it still may be possible to do so at lower concentrations than can be detected by PCR. On the other hand, in biofilm samples 'picking' becomes even more difficult to do and so PCR may be a more useful method. Further concentration of the environmental samples may improve the quantity of samples found positive by effectively lowering the detection limit(s) of the procedure(s). Although implementation of IMS as a part of this study was only partially successful, it may eventually prove to be an efficient means of doing this step without adding to the problems of contaminating bacteria. Whether the antibody used here was the most appropriate remains to be seen.

Although quantitative data were attempted in this study where possible, it is clear that the numbers recovered were generally low and cannot be treated with much confidence. They are likely to be underestimates, partly for the methodological reasons mentioned above, and also because legionellae are often regarded as being mostly in a viable but non-cultivable state. Where high numbers of legionellae are to be found, perhaps institutional water supplies, it may be possible to target a purer population rather than opting, as we did in this study, for maximum cell recovery with some concomitant loss of specificity.

Legionella were isolated from groundwater wells exhibiting very different water temperatures (8 to 45°C) and pH values (7.3 to 8.2) and thus the risks from legionellae in groundwater may be more widespread than previously appreciated. It is interesting to note from Figure 4.1, although by no means definitive, that at 30°C and above there were no negative *Legionella* samples. In in vitro culture, legionellae are fastidious with regard to pH but in the samples studied they were found over a wider pH range.

Water in general and groundwaters in particular are hidden resources. At one time, groundwater purity and availability were taken for granted on the assumption that particular hydrogeological patterns surrounding the groundwater wells did not allow microbial transport to water. Studies focusing on groundwater systems are now showing that chemical and/or microbial contamination of these systems can represent a risk to human health.

Survival of bacterial pathogens, such as *Legionella,* in soils and surface waters is recognized. Their transport to groundwater occurs during recharge through natural or anthropogenic heterogeneities in overburden and/or underlying bedrock. Preferential flow channels from drying of soils, presence of roots or animal and insect burrows, fractured bedrock, a water table close to the surface, and certain hydrogeological/geochemical features surrounding or upstream of a wellfield can all contribute to and potentially enhance such microbial transport. Considerably more attention is being paid to groundwater now that its vulnerability is recognized, however, the phenomenon of microbial transport remains poorly understood. It is also likely that high demands on a well can influence the migration of water within the aquifer, though this again remains poorly understood.

The main objective of this study was to investigate the level of contamination of groundwater systems by hydrotelluric bacteria of the genus *Legionella*. No field studies have been undertaken to evaluate *Legionella* transport to groundwater because until now these bacteria are primarily known to be surface water-contaminants. They are very often isolated from anthropogenic aquatic systems as well as lakes, rivers and ponds, suggesting their ubiquity in the water environment. However, given their close relationship with protozoan hosts, it is likely that legionellae can colonize anywhere that their protozoan hosts can thrive. This is supported by a bank filtration study (Schwartz et al. 1998) which suggests that legionellae were among the pioneer bacteria in naturally developing biofilms.

Groundwater and surface waters are interconnected. The presence of *Legionella* in groundwater systems under the direct influence of surface water is therefore considered logical. During this study, the focus was not on such systems, because water extracted from such systems is generally considered safe for human health. Most of the wells investigated were reported to be located in areas where they were protected by layers of clay, dense rock or other materials with very low permeability. Before this study and to our knowledge, no evidence for *Legionella* colonization of groundwater systems that are not known to be under the direct influence of surface water was available. However, virtually no continuously producing wells are likely to be completely free from some surface influences.

During this study *Legionella* were isolated from groundwater wells. Both bulk water samples and collected biofilms tested positive for the presence of *Legionella* in some wells. Overall, the average concentration of *Legionella* found in the bulk water was 3.5×10^2 CFU/L with a range from 25 CFU/L to 4.6×10^3 CFU/L. The biofilm results are expressed either as a density (when the surface area of the pipe section or water meter could be measured) or as a concentration when the surface area scraped was unknown. These results cannot be compared directly so they are presented separately. The average density of *Legionella* in the biofilm was 16.9 CFU/cm^2 with a range of 1-118 CFU/cm^2. For biofilm samples that were received suspended in water, the average concentration was 965 CFU/L with a range of 100 to 6.2×10^3 CFU/L.

When focusing on the results from water utility "A," it is clear that not only are *Legionella* present in their ground water systems, but also that the concentration and types of legionellae in the system seem to fluctuate from month to month. Prior to the beginning of Phase II of this study, it was found that PW_103 well was positive and PW_136 well was negative for the presence of *Legionella*. During Phase II, *Legionella* were isolated from the first 2 samples received from PW_103. These samples were collected during April and May 2000. The next 6 samples received from this well showed that *Legionella* concentrations had dropped to below the detection limits of the method. The last 4 sampling events for PW_103, during March, April, May, and June 2001, once again showed the presence of *Legionella* in the groundwaters. These results could indicate a possible seasonal distribution of *Legionella* in the groundwater system. However, the assumption of seasonal distribution in the groundwater system is not supported when examining the results from the second well from water utility "A." For the first three months of the study (April to June 2000), PW_136 was tested but due to technical problems was replaced with another well.

PW_138 was selected as the replacement well. When looking at the results, this well almost consistently tested positive for *Legionella* although the types of *Legionella* did vary from month to month. Since legionellae were recovered at similar levels over the 12-month period, this does not support the theory of a seasonal distribution. One surprising finding was the isolation of LLAP-like strains using cultivation without additional supplements since LLAPs are considered as not-yet cultivable species under standard conditions. The LLAPs that were isolated were from plates incubated at 30°C. This lower temperature may be more favorable for their growth than the conventional incubation at 35°C. The possibility that the isolate was misidentified is improbable since the sequence was 98% identical to LLAP-2, LLAP-9, *L. rowbothamii* (formerly LLAP-6), and *L. lytica*.

Newsome et al. (1998) found that a LLAP (identified as LLAP-14) was cultivable on BCYE differential and selective agars at room temperature and 30°C but not at 37°C. This finding adds support to our previous conclusion of 30°C being a more favourable growing temperature for some species assimilated to LLAPs. A 30°C growth temperature is also closer to the environmental temperature so the organism may experience less thermal shock and therefore increase its survival rate. Fluctuations in the levels and types of legionellae detected could also follow natural fluctuations in populations of their relative protozoan hosts.

In addition to groundwater well samples, water utility "A" was kind enough to provide additional types of samples for *Legionella* testing. KII RW and KII IFS samples were also taken from the raw water before any additional treatment but not directly from the groundwater well. Samples 05B, 15A, 21A, 32G, 72F and 94A were water meters taken from various locations that are fed by the same groundwater well field.

Although outside this study, samples taken from cooling towers fed by the same water distribution system of water utility "A" were also investigated. Cooling towers are often implicated as the source of community acquired Legionnaires' disease. These additional samples give an indication as to the wide spread presence of *Legionella* in the distribution system. The concentrations of the bacterium in the distribution system are, in general, 10 to 1000 fold greater than that found in the water and biofilm samples from groundwaters (data not shown). This was expected since anthropogenic systems act as natural amplifiers for *Legionella* and are often implicated in sporadic outbreaks of Legionnaires' disease. These results seem to indicate that fewer types of *Legionella* are present in the distribution system although, because of the lower number of samples examined, this difference could be artificial.

The *Legionella* concentration (specifically *L. pneumophila*) considered necessary to cause one or more sporadic cases of Legionellosis per year is 10^2 to 10^4 CFU per L (Ezzedine et al. 1989). Other studies indicate that *Legionella* only appeared to present a health risk to humans when a threshold value estimated at 10^4 to 10^5 CFU/L is reached (Leoni and Legnani, 2001). It should be noted that although *L. pneumophila* was isolated from some of the groundwater well samples, it doesn't appear to be the predominant species in raw groundwater. However, *L. pneumophila* appeared to be the predominant species present in the cooling tower samples. The concentrations in the cooling tower and water distribution samples were in the range considered to be a potential health risk. When considering all *Legionella* species, some of the groundwater samples contained concentrations between the suggested 10^2 to 10^4 CFU/L necessary for sporadic cases of Legionellosis.

Assessment of the risk of Legionnaires' disease outbreaks based solely on counts of *Legionella* may be misleading. In one study, results taken from the same cooling tower were statistically unrelated when samples taken a week apart were compared (Bentham 2000). Previous work has already demonstrated that *Legionella* concentrations in a system may vary as much as 3 orders of magnitude within 10 minutes (Bentham and Broadbent 1993). To what extent such fluctuating numbers represent changes in the planktonic population or result from shedding of clumps from biofilms and or dying protozoa, are unknown. Moreover, it is clear that the role of protozoa will also be important in quantifying risk of acquiring legionellae from contaminated sources.

All other water utilities that participated in this study submitted single water and biofilm samples from each well that was included by the utility. Depending on the water utility, either one or two wells were sampled.

The results show that *Legionella* were present irrespective of the water temperature since *Legionella* were present in the cold water samples received from Ontario, New Jersey, Nebraska and Michigan as well as in the warm water samples received from the Florida Keys, Texas, Florida, and Arizona. For this study, cold groundwater was defined as temperatures below 15°C. Because of the survey nature of much of the data and the following factors, the data obtained can be considered to be a minimum in terms of the numbers and types of legionellae present in the groundwater wells tested:

- Numbers and types of legionellae can fluctuate greatly on a temporal, and possibly seasonal basis – the data represent only a snapshot in time for each utility. Samples were collected at the convenience of the participating water utility and arrived at different times of the year, and both the water matrices and temperatures, as well as biofilms were quite different
- No single foolproof method for cultivation of all *Legionella* species exists, and the need to eliminate competing bacteria by a variety of treatment steps may also have eliminated some of the legionellae present originally in the samples.

It should be noted that the length of time allowed for biofilm growth on the PVC pipe sections received from water utility "A" was increased from 7 days to 15 days, and eventually to 60 days. It was not known whether *Legionella* were initial colonizing bacteria of biofilms so it was then decided to join two pipe sections in series to allow biofilm to develop for 60 days but to still be able to sample the wells each month. The PVC pipe sections received from all other water utilities allowed 30 days of biofilm growth on the pipe sections except for Water Utility F where only 15 days growth was allowed.

It was observed during cultivation of the water and biofilm samples from Texas (designated CS2 and CS5) that the samples had few other types of contaminating bacteria. The groundwater temperatures for these sample was >40°C. Temperatures between 45°C and 55°C are not optimal for the growth of Legionellae but give them a selective advantage over other bacteria commonly found in drinking water. This temperature may have acted as a selective decontamination procedure and favored *Legionella* over many other types of bacteria.

It has been found that legionellae are only ubiquitous in fresh water with conductivity between 10 and 120 $\mu S\ cm^{-1}$ (Heller et al. 1998). This is equivalent to 1 to 12 $mS\ m^{-1}$. The results indicate that although the conductivity is out of the range for ubiquity according to previous research, *Legionella* species were isolated from these samples.

Total organic carbon and turbidity have previously been positively associated with growth of *Legionella* sp. When looking at the samples from water utility "A," no conclusions can be satisfactorily drawn since the turbidity did not fluctuate with the absence or presence of positive colonies. There has also been a positive correlation of zinc with the ability to support *Legionella* spp. growth (States et al. 1987) although in this study zinc concentrations were not measured. Findings by another group of researchers indicated that there was no apparent correlation between the physical parameters of the water samples and the presence of *Legionella* (Lye et al. 1997).

Although questionnaire data was incomplete in many cases, a second stage statistical analysis was undertaken to try to find correlation or trends between some field parameters collected or selected hydrogeological data with the presence of *Legionella*. Statistical analysis was conducted by Dr. David Abrial (University of Saint Etienne, France) using ADE 4.0 ("Analyse de Données Ecologiques," ecological data analysis): a multivariate analysis and graphical display software (Thioulouse, et al. 1997), ADE-4: a multivariate analysis and graphical display software. *Statistics and Computing*, 7: 75-83). For the parameters investigated and using simple variance testing and discriminative factorial analysis, no indicator of the presence of *Legionella* has been identified.

In order to make the conventional cultivation technique as specific as possible, selective media and treatments were used during cultivation. For each sample received, a total of 36 plates were inoculated with the environmental sample of interest. Using this method, 15.9% of the plates inoculated were overgrown with other microorganisms. When comparing media by looking at the types of plates that were overgrown, 15.4% were CCVC plates, 66.1% were BCYEα plates, and 18.2% were GVP plates. These numbers are expected since BCYEα does not contain any of the selective ingredients present in GVP or CCVC. If the types of treatments are compared instead of the types of media, 37.5% of the overgrown plates were the result of direct plating, 31.3% were from acid treatment, and 30.5% were from heat treatment. These results indicate that the selective treatments were no more successful at eliminating contaminating bacteria than plating the sample directly onto the media. The inoculated plates were also incubated at two different temperatures to try and isolate a variety of species of *Legionella*. With regards to overgrown plates, 47.9% were from plates grown at 30°C and 52.1% were plates grown at 35°C; this was not a significant difference.

The conventional cultivation results were also analyzed to determine which media and treatments were most effective for isolating *Legionella* from the environmental samples. Of all the plates inoculated with samples, only 4.4% were found positive for *Legionella*. The percentage was calculated using all of the samples received during this study. This number is not necessarily reflective of the effectiveness of the media or treatments used since some of the samples included in the calculation may not have had *Legionella* present or the *Legionella* was at concentrations below the detection limit of cultivation. This result does give an indication as to the degree of difficulty and the time consuming nature of isolating *Legionella* using conventional cultivation. The effectiveness of the selective media and treatments were evaluated by comparing the treatments and medias that were found positive for *Legionella* growth. When looking at only the 4.4% of positive plates, 41.3% of colonies were isolated from CCVC, 22.8% from BCYEα, and 35.3% were isolated from GVP. It was expected that selective media (CCVC and GVP) would be more favorable for *Legionella* growth since the numbers and types of contaminating bacteria present is lowered. Less contaminating bacteria not only makes the *Legionella* colonies easier to identify since the chances of their being overgrown is lowered, but also inhibitory effects on *Legionella*'s growth by the presence of other bacteria would be reduced.

When comparing the selective treatments for *Legionella* isolation, 29.3% of *Legionella* isolates were from direct plating, 38.3% were from acid treatment, and 32.3% were the result of heat treatment. As with the numbers of overgrown plates, there does not appear to be a significant difference in the treatments used for isolation from environmental samples. Experiments conducted by another research groups have shown that the highest frequency of positive samples was obtained by inoculating the sample after filter concentration and heat decontamination (Leoni and Legnani, 2001). No significant differences were found in the isolation frequencies between their direct inoculum technique and the technique involving concentration and acid decontamination although their acid decontamination differed in that it was only applied for 5 minutes whereas in this study 15 minutes at pH 2.2 was employed (Leoni and Legnani, 2001). This suggests that the longer acid treatment time was more effective in isolating *Legionella*, probably due to the reduction of contaminating bacteria. The heat treatments used in both studies were identical.

One finding that was surprising was that 59.3% of the *Legionella* were isolated from plates incubated at 30°C and the remaining 40.7% were isolated at 35°C. Most laboratories attempting to cultivate *Legionella* use only 35°C. These findings indicate that more species and perhaps different species could be isolated using the lower temperature.

CHAPTER 5
SUMMARY AND OVERALL DISCUSSION (BOTH PHASES)

The number of wells that were positive for *Legionella* by any method are compared in both phases of the study in Table 5.1. It can clearly be seen that legionellae are widespread in groundwaters, though *L. pneumophila* were only found in a few samples.

The project was designed as a two-phase study. During the Phase I pilot study, the purpose was to demonstrate that *Legionella* could be isolated from groundwater samples using the methodological approach selected. The objectives outlined for this phase were achieved using twelve wells from two utilities and a QA plan to ensure data quality. All wells used groundwater that was considered 'warm.' Interestingly, culture methods picked up all 11 positive wells, whereas PCR showed only 4 wells positive.

In Phase II, the results were more evenly split between culture and PCR although culture still predominated for water samples and PCR may have had a slight edge for biofilm samples. In Phase II only 7 of 27 wells were negative by any of the methods applied. It can therefore be concluded that legionellae are widespread in groundwaters. Whether any of the negative wells would have tested positive on subsequent samples is unknown, but wells that were positive at Water Utility A and were tested monthly often produced negative samples indicating variability in both numbers and types of legionellae. In Phase II, both 'warm' and 'cold' groundwaters were investigated with legionellae found at all temperature and pH ranges. Legionella were found in shallow wells and also in those that were well over 1000 feet deep. It is not known whether any of the wells thought not to be under the influence of surface waters, are in fact impacted from unrecognized surface soils or waters. Clearly this deserves more investigation.

Attempts to identify definite indicators of *Legionella* occurrence or absence were mostly unsuccessful, although there are some weak positive and negative associations with certain metal species that may deserve further investigation.

An IMS system was designed to improve the detection of legionellae as an add-on to conventional culture and PCR methods. Although some success was achieved in the laboratory with reference strains, its use in the field did not offer any advantages over the conventional methods used in this study. The fact that IMS-culture and IMS-PCR each picked up 1 additional positive sample that had not been detected by the conventional techniques can be attributed in the one case to the sample not having been tested by conventional culture, and in the other case likely to additional testing but possibly to the numbers in the sample being below the detection limit of regular PCR. However, the potential importance of an IMS system in being able to lower the detection threshold suggests that more work in this area may still be valuable.

Table 5.1
Wells and utilities positive and negative for legionellae in this study

	Phase I	Phase II
Number of wells positive by any method	11	27
Number of wells negative by all methods	12	7
% positive wells	92	79
Utilities with all negative samples	0/2	2/16

Table 5.2
States participating in this study and the estimated percentage of population in them using groundwater as tap water

States	Arizona	Florida	Illinois	Michigan	Missouri	Nebraska	New Jersey	Texas
Percentage of population relying on groundwater	60	93	33	46	92	87	53	45

Source: Sustainability of Ground Water Resources *(1999), circular #1186, U.S. Geological Survey, Denver, CO.*

The limited work on protozoa in field samples from groundwater precluded any analysis but it is suggested that this deserves further and more systematic investigations. Some protozoa able to support *Legionella* growth were found, but additional species found should also be investigated for their ability to support *Legionella* survival and growth.

Samples were collected from various locations within the U.S. and Canada. Many water utilities were contacted to get their interest and approval for participating in this study, however. relatively few responses to our requests for participation were obtained. Whenever possible utilities located within states for which groundwater is considered the main water source were contacted. We are unsure of the reasons for the poor response, perhaps utilities are not associating *Legionella* contamination with groundwater and do not see the need to participate in such studies, or perhaps there is still an attitude that inhibits knowledge acquisition about their groundwater contaminants with the rationale that knowledge requires actions and costs.

Phase I samples were all from Florida, whereas in Phase II a year long monthly survey of one well from one of the original Florida utilities was continued and other sites from Arizona, Texas, Florida, Illinois, Nebraska, New Brunswick (Canada), New Jersey, Michigan, Missouri, and Ontario (Canada) were also included. Because water temperature was initially considered as the main parameter to investigate (as a possible indicator) during the study, a correct balance for water utilities enrollment between cold groundwater systems and warm-groundwater systems was necessary and respected. Cold groundwater systems were represented by water utilities from Illinois, Nebraska, New Brunswick (Canada), New Jersey, Michigan, and Ontario (Canada). Warm groundwater systems were located in Arizona, Texas, Florida, Florida Keys, and Missouri. Although it is difficult to assess whether the study significantly represented locations where groundwater use was high, some data for the percentage of people relying on groundwater in participating states is included in Table 5.2.

As far as we are aware, this is the first study that has unambiguously proven that *Legionella* constitute a part of the microflora of groundwaters not known to be under the direct influence of surface waters. Previous investigations in this regard were either very limited in scope or had used methodologies that were subsequently proven to be deficient (Lye et al. 1997; Brown et al. 1997). This leads to the conclusion that untreated groundwaters may disseminate *Legionella* to potable water distribution systems and may pose a risk for human health when considering the relatively high concentrations of the bacteria detected in this study. However, before deciding that disinfection needs to be undertaken
for such wells, it would be important to know that this did not exacerbate problems with *Legionella* and other important opportunistic pathogens that are resistant to disinfection such as

Mycobacterium avium. Source protection may also need to be implemented, but if surface influences on groundwater wells are unknown and perhaps distant, then this becomes a difficult proposition. Clearly, a better understanding of *Legionella* ecology in groundwater wells, as well as within treatment and distribution systems, will provide a knowledge base upon which policies can be developed and decisions can be based.

As reported above, samples of chlorinated water from some of the same production wells gave higher concentrations than the wells themselves - *Legionella* concentrations of 10^4-10^5 CFU/L.

SAFETY OF GROUNDWATER FOR DIRECT CONSUMPTION (UNTREATED GROUNDWATER AS A WATER SOURCE)

It is assumed that groundwater is safe for consumption without treatment because the soil acts as a filter to remove contaminants and/or because hydrogeological conditions do not favour microbial contaminations (*groundwater source not under the direct influence of surface water*), therefore many wells generally do not receive treatment. It is estimated by the U.S. EPA that 72% of the public water supply systems in the United States that use groundwater do not disinfect (see Figure 1.1).

During this study we have shown that *Legionella* are present in both treated and untreated groundwater. Knowing the ability of legionellae to colonize and be amplified in distribution systems, there may be risks from *Legionella* in distributing untreated groundwater. On the other hand, legionellae are relatively resistant to disinfection and could be selected for in systems with fewer competitors. Therefore, before deciding on whether and how to disinfect for this pathogen, careful consideration must be given to all the factors that may impact this decision. This includes a better understanding of *Legionella* ecology in groundwaters.

CHAPTER 6
IMPLICATIONS AND RECOMMENDATIONS FOR WATER UTILITIES

Legionellae, common in surface waters, are opportunistic human pathogens. Inhalation of the bacteria can lead to Legionnaires' disease (LD), a community- or hospital-acquired pneumonia, or Pontiac fever, a milder lung infection. The presence of legionellae in the U.S. is regulated through the Surface Water Treatment Rule (SWTR) promulgated by EPA. The SWTR also applies to groundwater sources known to be under the direct influence of surface waters. Thus, at the moment, only surface water systems and systems using groundwater under the direct influence of surface water are required to disinfect their water supplies.

In the U.S. and Canada, over 106 million residents rely on groundwater sources for their domestic water needs. Currently, EPA is seeking additional evidence for the occurrence of *Legionella* in groundwater before considering them for regulations. EPA is proposing, for groundwater sources that are not under the direct influence of surface water and for any system that mixes surface and groundwater (if the groundwater is added directly to the distribution system without prior treatment), a Groundwater Rule (GWR) that will include a Groundwater Disinfection Rule intended to set up maximum contaminant level goals (MCLGs) and minimum levels of disinfection for numerous pathogens.

Until now there has been a lack of evidence for the presence of *Legionella* in groundwater sources that are not recognized as under the influence of surface water. Based on the results of this study, it is legitimate to ask whether groundwater could be a risk when considering legionellae. Although this is still difficult to answer definitively, this study provides strong evidence that there might be a risk associated with the use of water from those groundwater systems that do not disinfect, even if that risk is mainly in seeding a distribution system where legionellae can subsequently multiply, and eventually the plumbing systems of end users. It needs to be reemphasized here that legionellae are not known to be infectious by ingestion, and so the risks from exposure to water contaminated with legionellae are to humans exposed to small water droplets that are inhaled. This could occur during showers, from humidifiers, in hot tubs, or any other functions where aerosols can be generated. The segments of the population most vulnerable to the severe consequences associated with LD are the elderly or those with immune systems compromised by disease or medication. However, due to the lack of knowledge regarding the incidence of Pontiac fever, it is not known how widespread infection with legionellae is, or what fraction of that is drinking water-related. Nevertheless, changing lifestyles, with perhaps more indoor time and exposure to systems that could harbour legionellae, suggest caution should be exercised by water utilities.

It also needs to be emphasized that relatively little is known about the infectious dose of legionellae by inhalation. Although relatively high numbers may be necessary to cause the severe pneumonia in most individuals, the dose for immunocompromised individuals is unknown. Other unknowns include the infectious dose for Pontiac fever and the influence of inhalation of amoebae containing legionellae. Water may be contaminated by *Legionella* and may never have been associated with clinical cases. This should not be a cause for complacency, because changes in the bacteria type, numbers and virulence, or in the susceptibility of the host(s), could still cause clinical infections.

There are a number of factors that can affect *Legionella* testing and the numbers and types of legionellae detected in water systems. These include:

- Bacteriological testing gives a '**snapshot**' only of the *Legionella* contamination of the sample tested. For water, this is usually only a minute fraction of the well output. Multiple samples and preferably a regular survey of a given system is likely to be more informative than point sampling events. During our study, only one sample from each well was tested for most utilities. In spite of this, a significant proportion of the utilities had legionellae. It is highly likely that this number would have increased if regular samples had been taken from all utilities. It is also obvious that one cannot consider a single well to be representative of a well field or an entire production system.
- A variety of difficulties, discussed earlier in the report, remain in cultivating legionellae from water and biofilms so **data will always tend to underestimate *Legionella* presence**. A word of caution here that water systems tested to be free of legionellae will not necessarily remain so. Many hospitals routinely use temporary hyperchlorination and increased temperatures to reduce the risks of legionellae for sensitive patients. Although legionellae may temporarily disappear, the systems are rapidly recolonized and require regular treatment.
- Within a given system, it is very likely that the **concentration of *Legionella* may vary** from one sampling event to another (for instance, biofilm may be released from wall pipes and suddenly increase the bacterial concentration in the sample). Variation could also occur based on natural fluctuations in host protozoan strains.

LEGIONELLA SPECIES ISOLATED FROM GROUNDWATER (OCCURRENCE AND DISTRIBUTION) VS. *LEGIONELLA* SPECIES PREVIOUSLY ASSOCIATED WITH DISEASE

Several *Legionella* species are not known to cause disease in humans. However the clinical significance of others may be under recognized because of difficulties in their cultivation. This study attempted detection of *Legionella* from a total of 114 samples of raw (most) and finished (few) waters as well as biofilms from underground systems taken at the well head (water temp. <15-45°C). Sixteen water utilities from eight U.S. states and two Canadian provinces participated. *Legionella* was unambiguously isolated in 33% of the groundwater samples tested.

To date, 48 species of *Legionella* have been described (see Table 1.1). Of these 48 species, 21 have been related to clinical cases and most of them with LD. Some evidence also suggests that species belonging to the "*Legionella*-like Amoebal Pathogens" grouping may also be responsible for LD (Marrie et al. 2001). Other species remain, strictly speaking, environmental isolates. Technical progress in *Legionella* speciation has shown that some uncommon (environmental) species may be responsible for pathology (Lo Presti et al. 1997; Lo Presti et *al,*. 2000) and suggesting that "environmental" species may also be pathogenic for humans when particular conditions are met. Therefore, we believe that with current knowledge all species should be considered potentially pathogenic.

Although our identification strategy is still imperfect (partial 16S rRNA gene sequencing), at least 11 *Legionella* species have been identified in the samples processed with some samples yielding more than one species. This suggests that a large distribution of *Legionella* species can be expected in groundwater systems. In this study, *L. pneumophila* (the most common causative agent of LD) was infrequently isolated, suggesting that groundwater systems may not normally be favorable for its growth. However, its detection as the predominant *Legionella* species in separately analyzed cooling tower samples from Water Utility A suggests that it could have been present at low numbers in the source water and then becomes selected for in a more favourable environment in the cooling towers.

LEGIONELLA CONCENTRATIONS (LEVEL OF CONTAMINATION) IN THE SAMPLES TESTED

Based on our cultivation process, *Legionella* were isolated from the samples tested from 10^2-10^5 CFU/L in water and up to 1.2 x 10^2 CFU/cm^2 in biofilms. Although not the primary purpose of this study, samples of chlorinated water (distribution system) from some production wells were found to contain *Legionella* concentrations at 10^4-10^5 CFU/L. Furthermore, when water samples from some cooling towers fed by the same distribution system were separately tested and, 10^4-10^5 CFU/L of the bacteria were detected.

There is no consensus about the critical level of *Legionella* contamination that should encourage a corrective action to be taken. However, as discussed earlier, some studies indicate that *Legionella* would appear to present a health risk to humans when a threshold value of 10^2 to 10^5 *Legionella* /L is reached.

Several countries (for instance, Switzerland, France) estimate that the *Legionella* contamination threshold, for which a corrective action has to be taken, should be correlated with host risk factors. It is therefore understood in these countries that the host sensitivity is the best criterion to draw guidelines for corrective actions. In France for instance, it is considered that for public facilities that do not provide health-care support, 10^3 *Legionella*/L is the critical threshold for a corrective action to be decided on. For Switzerland and France, in hospitals, high-risks units (intensive care, etc.) should ensure that their level of contamination does not exceed 0.1 CFU/mL (10^2 CFU/L). In other situations, a *Legionella* concentration of 1 CFU/mL (10^3 CFU/L) is the threshold necessary to implement a decontamination procedure.

In France again, although still recommended to be 10^2 *Legionella*/L, it may be soon recognized that, within high-risk associated environments (such as hospitals, health-care treatment facilities), the MCLG in the distribution system should be lower on a global basis (50 *Legionella*/L, i.e. the detection threshold of the cultivation assay). Again, these regulations are taking into consideration risk factors (mainly their immunological status) associated with potential hosts. For the U.S. Centers for Disease Control and Prevention (CDC), corrective actions need to be taken when more than 30% of the samples yield positive results.

Our results were occasionally high enough to trigger corrective actions in some jurisdictions. This suggests that distributing groundwater from wells that are not under the direct influence of surface water directly to consumers without any treatment may lead to continual *Legionella* seeding and spreading within distribution systems, and present potential hazards for human health. Thus, water systems providing untreated groundwater must be considered at risk for *Legionella* irrespective of the hydrogeological pattern surrounding the well(s) or their sensitivity to fecal contamination. However, given the relative insensitivity of legionellae to disinfection by chlorine and their ability to remain viable within protozoa and their cysts, *Legionella* contamination of distribution systems may still occur if chlorination alone is used. More work is clearly needed to understand the influence of varying disinfection technologies on legionellae, particularly within distribution systems and because of their complex association with protozoa.

Our findings provide new data on the occurrence of legionellae in well waters. Based on the level of contamination observed and the general assumption that all *Legionella* species should be considered pathogenic, we therefore recommend that *Legionella* be considered by the U.S. EPA and other similar jurisdictions for regulation in any future rules on groundwater quality. It would be particularly recommended to set up MCLGs for *Legionella* depending on the risks associated with potential consumers. This would be more meaningful than defining a strategy towards *Legionella* based solely on the presence of one particular *Legionella* species in a colonized system. A logical approach would be to define risk classes for facilities fed by untreated groundwater considering the health status of the consuming population. For instance, health-care facilities and hospitals would be recognized at the highest risk class and a MCLG could be proposed by competent jurisdictions according to their understanding of the *Legionella* risk in humans. Implementation of such a classification has the potential to reduce *Legionella*-associated deaths.

Accurate estimates for the number of cases of LD in North America are not currently available, and there is virtually no information on the potentially more common but self limiting Pontiac fever. The impact of this disease in our communities is likely to increase substantially with the numbers of the elderly, debilitated and immunocompromised individuals. We, therefore, consider it important to institute appropriate measures to reduce the risk to human health from this potentially serious opportunistic pathogen.

TECHNICAL COMMENTS AND RECOMMENDATIONS

Although this study provided definite isolations of legionellae from groundwaters using culture as the gold standard, the method used was relatively complex and the skill level needed was relatively high. There is still, therefore, a need to simplify and standardize the methods for detecting legionellae, especially to simplify routine testing of sources, such as groundwater, where their concentrations are relatively low.

It is recommended that individual utilities providing untreated groundwater acquire an understanding of the potential for their system(s) to be contaminated by legionellae, both from source water and growth within the distribution system. For source water, this may range from introducing scheduled testing at relevant locations to re-evaluating the potential for unknown surface water influences from 'upstream' sources. Most unconfined aquifers receive some contamination from the surface during recharge, even if not deemed to be directly under the influence of surface sources. However, because contaminating surface sources may be some distance from the wellhead, this can be difficult to determine and to control. Contaminating sources may even be in a different political jurisdiction. Nevertheless, attempts to understand watershed hydrogeology and potential contaminant sources should be stepped up. As development pressures and intensive agriculture increases, best management practices (BMPs) for well design and installation, as well as for land use and waste disposal to land are evolving to protect both surface and groundwater sources. Application of BMPs should be stringently adhered to in the interests of public health for such measures will not only protect against legionellae but many other pathogens deposited into/onto or naturally occurring in surface sources. Amplification of legionellae within distribution systems may make them easier to detect at predictable points in distribution systems where water quality is generally poorer. With appropriate cooperation, it may also be possible to use data from sentinel locations such as hospitals to monitor the distribution system.

The drinking water industry is unaccustomed to dealing with respiratory pathogens, and usually performs direct analysis on water or biofilms to detect either pathogens acquired by ingestion, or their microbiological indicators. To get a better estimate of user exposure to legionellae and other respiratory pathogens such as non-tuberculous mycobacteria that may be present in drinking water biofilms, it may be desirable to consider an entirely different sampling strategy. This could involve the collection of water sprayed from showers, for example, but it would need to be performed at 'simulated consumer taps/showers' in order to represent the usual daily exposure that may occur.

It may also be desirable for the industry as a whole to have a better understanding of the ecology of legionellae. Sampling for legionellae is complex and it would be desirable to be able to test for a more simply detected surrogate indicator. Currently no such indicator is known or recognized. The data from this study do not support any definitive physicochemical indicator(s) and much work would be needed before any bacterial or protozoan indicator(s) could be proposed. However, improved understanding of *Legionella* ecology in groundwater and associated distribution infrastructure could lead to ultimate selection of such an indicator.

Ideally there should be some national co-ordination of efforts on legionellae so that a database can be built up, or the information added to a preexisting database.

Although no infectious dose of *Legionella* can be easily determined, mainly because host risk factors may be a lot more important that the virulence of the species or the strain considered, we recommend consideration of an MCLG for legionellae in untreated groundwater. The World Health Organization has previously suggested a threshold of 1000 *Legionella* per L, provided that the groundwater source does not feed any health-care facilities. However, the current difficulties of quantitative culture of legionellae from natural water samples and lack of a standardized method may postpone such a decision. If this MCLG is exceeded, disinfection practices would need to be initiated. In this context, it is possible that chloramination would be a better option than chlorination for the control of *Legionella* in distribution systems (Kool, Carpenter, and Fields, 1999). However, further controlled studies are needed to confirm this initial observation in a variety of scenarios. In general, eliminating contaminants in water supplies should be secondary to the need to protect the water supply from the initial source of the contamination.

If necessary, the utility must be prepared to communicate factual information to its consumers in a simple, reasoned, open and straightforward manner. Four fact sheets on *Legionella* are produced by the U.S. EPA to assist utilities and their customers:

http://www.epa.gov/waterscience/humanhealth/microbial/legionellafs.pdf
http://www.epa.gov/waterscience/humanhealth/microbial/legionellachild.pdf
http://www.epa.gov/waterscience/humanhealth/microbial/legionellaha.pdf
http://www.epa.gov/waterscience/humanhealth/microbial/legionella.pdf

CHAPTER 7
FURTHER RESEARCH NEEDED

The following are specific or more general research needs that have been identified based on the findings of this investigation. They are not in any particular order of importance.

- Higher concentrations of certain metals, which were only weakly associated with *Legionella* presence or absence in this study, deserve further systematic investigation. However, such an investigation should be carried out in association with other factors known to promote legionellae in water systems, such as protozoa. If there is an effect of metal ions, is it on the bacterium or its protozoan host(s), or on both?
- Further improvement of the *Legionella* IMS methodology.
- Also, in the above, or in separate studies, it would be valuable to better understand the role that protozoa play in the persistence of these organisms in soils and in their transport to groundwater. The role of encysted forms of amoebae would be particularly important to study in view of the depths of some of the wells that were found to contain legionellae and the implication for long term environmental persistence.
- Due to the relatively high resistance of legionellae to disinfection, and their ability to 'hide' from disinfectants in protozoa, it is important to understand the relative roles that drinking water disinfectants can play in controlling or enhancing the presence of legionellae in drinking water treatment and distribution systems. Several questions arise in this regard – the direct role of the disinfectants on the legionellae; - the role of the disinfectants on the protozoan host(s); and how readily particular *Legionella* strains can adapt from one protozoan host to another? If the normal chlorine-based disinfectant residuals have differential effects based on laboratory studies, can this be supported by field observations? Are legionellae selected for as a result of removing some of the competition from those bacteria which are more susceptible to disinfection? Do chlorination or chloramination have any effects on *Legionella* selection, particular on *Legionella* species or strains that might be more adapted and/or more virulent or pathogenic for humans? What are the impacts of a switch from chlorination to chloramination on *Legionella* and other members of microbial communities?
- Legionellae are part of a whole community of opportunistic pathogens that interact with protozoa, that can be found in drinking water, and whose presence is potentially affected favourably by the presence of disinfectant residuals. Legionellae should therefore be investigated in parallel with mycobacteria, *Helicobacter* and possibly *Campylobacter*. More research is needed to understand the ecology of these pathogens in drinking water sources (ground and surface waters) and infrastructure.
- Clearly there is also a need from the point of view of individual water utilities producing untreated groundwater to better understand the influences that there may be on the production wells. There is a tendency to assume that distance from obvious sources of contamination and depth of wells protects. Although this is partially true,

for pathogens that can persist in protozoan cysts for prolonged periods, distance and depth may be no barrier.

- Aerosols: study potential relationships between the concentration of *Legionella* in water and their concentration in aerosols in the presence and absence of their protozoan hosts; evaluation of the survival of *Legionella* within aerosols; study relationships between aerosol exposure and legionellosis).

APPENDIX A
TECHNICAL DIFFICULTIES AND TECHNICAL NOTES

SAMPLING DIFFICULTIES

Since Water Utility A provided us with samples on an ongoing basis throughout the study, most of the technical difficulties related to sampling occurred in their well fields. On June 15th 2000, during the third sampling of Phase II, we were informed that all of the wells of "Water Utility A" were shut down for a period of 4 hours to switch the pumps over to diesel power - electrical problems were encountered in the wellfield. It is not known if this had any impact on the June 19th samples from the pipe sections, but because the well was shut down and could not be restarted, no water samples for analysis could be collected. Therefore, we received the pipe fitting for PW_136 and no other parameter data for this well were obtained.

On June 20th 2000, we were informed that the PW_136 well pump went into phase loss and shut down; the motor received a direct hit from a lightening strike. As a consequence, no samples could be obtained from PW_136 for an undefined period of time. Therefore, we decided in conjunction with people from "Water Utility A" to chose another well for the continuation of the study and to go back to PW_136 as soon as it would be back on line. Only one well found positive for legionellae during Phase I, was back on line and available for sampling, and this (PW_138) was chosen to temporarily replace PW_136 for the survey.

PW_138 is located at the northeastern part of the well field in an open field. It was installed in 1966. The casing depth is 58(ft). Total depth is 778 (ft). The annual production (million gallons) (1988-1994) ranges from a high of 154.59 (1993) to a low of 34.20 (1994). The monthly average (million gallons) (1988-1994) ranges from a high of 2.88 (1993) to a low of 2.85 (1994). The water temperature is roughly constant at approximately 25°C. It is a low volume well (0.2 MG/day) that has a troubled history with MMO-MUG. Of 42 samples, 23 had total coliform presence. *Enterobacter cloacae* are the predominant isolates. *Pseudomonas fluorescens* was cultured from a total coliform negative test that was turbid. The Biological Activity Reaction Tests (BART™) indicated aerobic, enteric, sulfate-reducing, pseudomonads and fluorescing pseudomonad bacteria present and is employed as a first-step measurement of microbial activity in the wellfield. BART™ was developed by D. Roy Cullimore (in *Practical Manual of Groundwater Microbiology,* 1992) for detecting the presence and aggressiveness of selected groups of potential nuisance bacteria.) Additionally, some important rains storms occurred during the May-June sampling period after a very dry year and the driest month of May in recorded history.

One potential problem highlighted by people from "Water Utility A" is that when one well shuts down, water from the other wells will continue to flow backwards through the well which is shut off due to the lack of pressure. This means that the negative well may become positive if it becomes contaminated with water from a positive well positioned on the main line. We do not know whether this affected our results.

On April 30th 2001, people from "Water Utility A" informed us that the wells are no longer in the manual position; they are now kept in the auto position. This means that the well pumps will turn on and off. In the "on" position water is flowing through the pipe from the designated well. When in the "off" position water still flows through the pipe sections but the water is coming from neighboring wells. This might have affected our samples from "April-May" and following sampling events.

Every time the sampler tried to tag it for manual operation, the tag got ripped off. Changes in the management of water utility "A" had occurred, the wellfield is owned by a different authority meaning that it was less easy to get control of the situation for the sampling events.

During this period, some plugs of sand were frequently observed during sampling within the pipe sections meaning that the well pumps were sometimes sucking up sand. There are 2 possibilities to explain this phenomenon: (a) the utility is over pumping (given the dry conditions encountered recently) or, (b) the well casing may be breaking down.

It has been proposed by the utility that we analyze biofilms obtained from water meters removed from the distribution system fed by the groundwater well field (finished water). Six water meters classified as consumption meters (meaning that 1 to 2 million gallons has passed through them) installed in customers' houses were removed and the biofilms obtained were analyzed for the presence of *Legionella*. Two different brands of meters were accessible (both made of brass, lead and copper with inner parts in plastic).

In June 2001, both of the wells (PW_103 and PW_138) were off and PW_103 was out of service. Pipe sections from both wells were removed and water samples were collected from PW_138 only (see the sampling paragraph for PW_103 "May-June samples). From this date, PW_103 was alternatively shut down due to pumping problems and no water sample from this well was obtained until the "August" sampling event. From August 2001, PW_103 was no longer available for sampling studies. Water utility "A" proposed to sample an additional well to finish the study. PW_116 was selected by water utility "A." This well will not be part of the survey process strictly speaking but will be tested anyway.

CULTIVATION

Viable plate count or most-probable-number techniques have been, and frequently still are, used for quantification of active cells in environmental samples. However, these methods only select for certain organisms (Kalmbach, Manz, and Szewzyk, 1997). For oligotrophic to mesotrophic aquatic samples, it has been frequently reported that direct microscopic counts exceed viable-cell counts by several orders of magnitude; the same is true for sediments and soils (Byrd, Xu, and Colwell, 1991; Steinert et al. 1997). This phenomenon is known as the "great plate count anomaly" described by Staley and Konopka (1985). This fact is explained by the possibility that some bacteria may enter a state in which they retain viability but cannot be cultured from the environment. This phenomenon has also been reported for *Legionella* (Hussong et al. 1987; Paszko-Kolva et al. 1992; Barer et al. 1993; Yamamoto, Hashimoto, and Takayuki, 1996; Steinert et al. 1997). The possibility that *Legionella* enter a viable-but-nonculturable (VBNC) state or a dormant state when exposed to low-nutrient environments cannot be excluded (Hussong et al. 1987; Paszko-Kolva, Shahamat, and Colwell, 1992; Steinert et al. 1997; Grimm et al. 1998) even if *L. pneumophila* can maintain a vegetative state (at least 600) when incubated in a low-nutrient environment (James et al. 1999). Therefore, it is tempting to speculate that some Legionellaceae enter into VBNC states in groundwater. This has been recently hypothesized by Lye et al. (1997).

Non-culturable *Legionella* species (because of their ability to grow almost exclusively in amoebae) such as *L. fallonii, L. drozanskii* and *L. rowbothamii* (Adeleke et al. 2001), formerly named *Legionella*-Like Amoebal Pathogens (LLAPs), are probably underestimated causes of human disease (Rowbotham, 1993; Benson et al. 1995; Butler et al. 1995; Abu Kwaik et al. 1998). Considering that approximately 50% of the 0.5 million annual cases of pneumonia in the United States of America are of unknown etiology, the LLAPs might be responsible for at least some of these cases (Abu Kwaik et al. 1998). Their occurrence in groundwater is also unknown and should probably be correlated with the presence of supporting protozoa.

Regardless of physiological state of *Legionella* cells (dormant, VBNC) or strict intracellular growth state (LLAPs), PCR should be a valuable tool to detect them in samples of groundwater. However, PCR detection has its own difficulties as outlined below. PCR can also be a valuable tool for *Legionella* identification from cultured colonies – see below.

An additional problem to consider, when bacterial identification is required, is the number of colonies requiring sub-typing after cultivation. No particular recommendation exists, but Horbach et al. have shown by studying 7 infections, and typing 10 to 20 *Legionella* colonies, that 3 of these infections were caused by multiple strains (Horbach, Naumann, and Fehrenbach, 1988). Harrison et al. recommend testing a number of colonies equal to the root of the number of grown colonies on the tested media (Harrison et al. 1990).

SEMI-NESTED PCR

For this study, the oligonucleotide primers were designed from the partial sequences of 16S rRNA genes of *Legionella* species (Miyamoto et al. 1997), which were obtained from the Ribosomal Database Project (Olsen et al. 1992).

For the first step of the semi-nested procedure, 45 µl of a master mix (see below) was added to each tube along with 5 µL of sample that contained the DNA to be amplified (extracted DNA from samples). Two oligonucleotides (named LEG 225 and LEG 858) were used as primers enclosing a 654-bp fragment of the 16S rRNA gene. LEG 225 (5'-AAGATTAGCCTGCGTCCGAT-3') was located at positions 225 to 244 and LEG 858 (5'-GTCAACTTATCGCGTTTGCT-3') was complementary to positions 880 to 859 (*E. coli* numbering). LEG 225 was first selected from a large database of the 16S rRNA gene (Miyamoto et al. 1997).

The second step of the semi-nested PCR used 45 µl of a master mix and 5 µL of the DNA product amplified in the first step. For the second-step PCR, LEG 448 (5'-GAGGGTTGATAGGTTAAGAGC-3') and LEG 858 were used as another set of primers enclosing a 430-bp fragment of the 16S rRNA gene (Miyamoto et al. 1997).

PCR master mix:

Master Mix Composition for a 10 µl sample of DNA (total volume 100 µl):

- 10 µl 10x PCR Buffer (10 mM Tris-HCl (pH 8.3), 50 mM KCl, and 0.1 mg/mL of gelatin)
- 6 µl (1.5 mM) of magnesium chloride
- 6 µl (1 mM) of forward primer
- 6 µl (1 mM) of reverse primer
- 16 µl dNTP (0,2mM dATP, dGTP, dCTP, dTTP)
- 45 µl ultra-pure water
- 1 µl *Taq* polymerase

To resolve the bands of interest from PCR, agarose gel electrophoresis was performed. Agarose was used at 1% (w/v) in Tris-Borate-EDTA (TBE) solution. The gel was then stained for 15 minutes in a 10-µg/mL solution of ethidium bromide, destained in 1X TBE for 15 minutes and placed on a UV transilluminator to resolve any bands. A picture of the gel was taken and data reported.

LEGIONELLA IDENTIFICATION USING PCR

The semi-nested PCR procedure described here is a *Legionella* genus-specific assay. It was also used for identification of presumptive colonies following cultivation. The PCR product from these colonies was visualized by agarose gel electrophoresis, confirming that the isolate was *Legionella.* The PCR product was run again on an agarose gel as per above-mentioned instructions. The band was then excised from the gel and the DNA was isolated using a gel purification kit (Qiagen). DNA sequencing of the amplified 16S rDNA fragment (first step of the semi-nested PCR assay only) was performed at the University of Ottawa Biotechnology Research Institute (UOBRI). The sequences obtained were compared with available sequences in the DDBJ, EMBL and GenBank databases with the Gapped BLAST program (Altschul et al. 1990) through the Internet (http://www.ncbi.nlm.nih.gov) to provide the most likely identities of the isolates. These results should be considered with some caution because partial 16S rDNA segment sequencing is not the ultimate tool to definitively assign a species name to strain (Fox, Wisotzkey, and Jurtshuk, 1992; Stackebrandt and Goebel 1994).

TECHNICAL ISSUES IN PCR AMPLIFICATION

PCR methodologies are hampered by numerous parameters when environmental samples are processed.

Initial DNA Extraction

The initial extraction of nucleic acids is a crucial step since, unfortunately, not all microorganisms lyse equally well (Picard et al. 1992). During Phase I, DNA extraction was done following published procedures (Jaulhac et al. 1992) mainly by the use of a chemical lysis buffer and/or by heat-thawing treatment of the samples. These treatments proved to be efficient in the past to extract DNA from contaminated clinical specimens as well as environmental samples (Jaulhac et al. 1992; Jaulhac et al. 1993). During Phase II, we decided to use a commercially available DNA extraction kit (Dneasy extraction kit, Qiagen) to try to increase the sensitivity of the PCR assay.

PCR Amplification

Through the use of a published protocol (Miyamoto et al. 1997), genus-specific 16S rRNA primers were used in order to selectively amplify DNA from *Legionella* (detection approach). In this case, detection is simplified but identification is still conditioned by sequencing amplicons. PCR was then used for two different purposes:

- As a detection method;
- As a tool to identify purified isolates obtained after cultivation.

For both purposes, several technical difficulties are known:

Preferential Amplification of Some Sequences

When using PCR as a detection method, one should keep in mind that a potential bias exists for the preferential amplification of certain templates over others (Reysenbach et al. 1992; Lee, Zo, and Kim, 1996). This could result in a misrepresentation of the material abundance of certain rRNA genes. It has been shown that such differential amplification of rRNA genes from mixtures with other rDNAs is not a simply theoretical problem. In the worst case, it can prevent the amplification of certain rDNAs (Reysenbach et al. 1992). Because of selective priming or higher-order structural elements, certain sequences might be discriminated against when compared with others. The number of copies of the targeted sequence is also an important parameter, which might explain the preferential amplification of some sequences versus others. In this context, rRNA genes are more likely to be amplified because of their high copy number (Okpara et al. 1996). Nevertheless, because environmental samples contain an unknown mixture of microorganisms and other substances, it is possible that legionellae that are present may not be detected even in the absence specific PCR inhibitors.

Inhibitory Substances

When using PCR as a detection method, numerous substances in various water samples have been reported to cause PCR inhibition (Wilson, 1997). PCR inhibition needs to be anticipated and no standardized methods exist to remove all of them from a sample even if efficient assays have already been published, and reviewed (Wilson, 1997)). Elimination of these interferences often requires dilutions of the water samples (Lye et al. 1997) with the risk of a proportionally decrease of sensitivity. Purification of DNA has been reported to be efficient to eliminate inhibitors. Some studies on *Legionella* (Maiwald et al. 1995) and mycobacteria indicate that extraction procedures employing multiple steps to purify DNA from samples had lower rates of PCR inhibitors than those with fewer steps. Chromatographic methods have been shown to efficiently eliminate PCR inhibitors consisting most likely of heavy metals and organic matter (humic compounds) in water samples (Catalan et al. 1997). Complex associations may lead to the formation of organometallic compounds which are more difficult to exclude.

Finally, in order to test an environmental sample for the presence of *Legionella* by a PCR-based method, one has to first test duplicate samples (repeatability) then, when inhibitory samples are suspected, numerous dilutions of the sample need to be processed. To confirm a potential inhibitory effect, adding pure DNA to the sample is also required. If we consider additional positive and negative controls, one should process up to 5 assays per sample. Screening large number of samples would then require numerous tests.

Difficulties in Sequence Retrieval

When using PCR as an identification tool, formation of chimeric rDNA sequences can occur. These chimeras, assembled from sequences of different species (Wang and Wang 1997), are generated when a DNA fragment of one gene anneals with a homologous template to prime the next cycle of DNA synthesis. Prematurely terminated DNA strands, especially in later cycles of PCR, have been shown to be the main cause of chimera formation (Suzuki and Giovannoni 1996). Two main factors will increase the probability of chimera formation:

a. The availability of partial-length rDNA fragments present in low-molecular-weight genome DNA preparations or generated by premature termination of elongation during PCR; and
b. The percentage of highly conserved stretches along the primary structure of rDNA types, where, after denaturation, single strands originating from different rDNAs can anneal in highly complementary regions (Wang and Wang 1997). This artifact may incorrectly be interpreted as an indication of additional biodiversity in the natural sample (Wang and Wang 1997). Whereas two distinct populations represent real diversity with two different rDNA types, the clone library might indicate the presence of a continuum of rDNA types with only small differences between the individual types.

Keeping in mind the number of steps involved in the retrieval of rDNA sequences, one should be careful to regard these sequence collections as unbiased reflections of the microbial communities living in the natural ecosystems.

16S rRNA Gene Sequencing As a Tool To Identify Organisms

Effective identity of 16S rRNA sequences is not necessarily a sufficient criterion to guarantee species identity (Fox, Wisotzkey, and Jurtshuk, 1992). DNA/DNA hybridization or 16S rRNA sequencing of 1,000 or more bases is the optimal methods for determining phylogenetic relationships above the species level (Murray et al. 1990). However, for Stackebrandt and Goebel, only complete sequences (16S rRNA) allow reliable phylogenetic comparison with the available database and caution must be taken when using 16S rRNA gene sequencing as a tool for identification considering that several groups of organisms have been identified sharing almost identical 16S rRNA sequences but in which DNA hybridization is significantly lower than 70% (Stackebrandt and Goebel 1994). It is considered that organisms exhibiting less that 97% sequence homology will not reassociate (DNA/DNA hybridization) to more than 60% (Stackebrandt and Goebel 1994). Hence, such isolates cannot be related at the species level.

As a consequence, the higher the percentage of homologies for 16S rRNA sequence, the better one can assign a species name to an isolate although even for 16S rRNA identity of 99%, DNA/DNA hybridization may or may not document the existence of species identity (Fox, Wisotzkey, and Jurtshuk, 1992). We recognize that our identification results must be considered with some caution because only 500 bp fragments have been analyzed.

IMMUNOMAGNETIC SEPARATION ASSAYS

As described previously, the recovery of bacteria by IMS may be hampered by several parameters. Ultimately, the reaction is always a product of the antibody(ies) available and the antigen-antibody reaction parameters. Clearly, a choice was needed between maximizing the capture of target cells and accepting a high contamination level, or obtaining a smaller population of pure target cells. Given the limited numbers of legionellae in our samples, our choice was to make sure whether the *Legionella* was present or not in the samples tested by maximizing cell recovery. As a consequence, the background contamination observed on some of the inoculated plates following IMS was high because few washing steps were used to avoid loss of legionellae to levels lower than the sensitivity of the assay. In order to increase the assay sensitivity, two procedures were tested (a combined procedure and an uncombined procedure) to find the better approach.

An additional possibility to explain the background contamination on the plates after IMS is the lack of specificity of the mouse monoclonal antibody (Mab known as 22/1, IgG1 subclass) used. This Mab is able to recognize a *Legionella*-specific epitope of the Mip (macrophage infectivity potentiator) protein (Helbig et al. 1995). Several Mip-like proteins have been described in various groups of bacteria, therefore the Mip epitope targeted on the surface of *Legionella* may share structural homologies with Mip-like epitopes from other bacteria; this makes a non-specific binding of the antibody possible.

In their study, Helbig et al. (1995) showed that there were no possible cross-reactions with non-*Legionella* bacteria using this Mab. However, it is always possible that some unknown or not previously encountered bacteria such as those found in environmental samples exhibit extensive homologies for some surface epitopes. Another possibility identified was the use of a sheep anti-mouse IgG1 coated Dynabeads® M-450 (Dynal Inc, NY, USA) as the secondary antibody to detect *Legionella*-Mab 22/1 complexes. It was obvious that this secondary antibody could possibility recognize other bacteria. These reasons encouraged us to use the two procedures described earlier (combined and uncombined). The combined procedure would ensure a better specificity of the assay.

BIOLOGICAL ACTIVITY REACTION TEST (BART™)

The Biological Activity Reaction Test (BART™) was employed as a first-step measurement of microbial activity in the wellfield (water utility "A"). BART was developed by D. Roy Cullimore for detecting the presence and aggressivity of selected groups of potential nuisance bacteria. The tubes are checked daily for one month for indicator reactions such as color changes, turbidity, sediment, precipitation, and colored formations such as rings and bubbles.

Two forms of data were obtained: of delay (days) and the reaction type (RX) which occurred. It can be interpreted that the shorter the days the greater the population size and activity level. At the same time, the RX gives an indication of the types of bacteria present in the water samples. Such days and RX data can be used to compare, over time, whether a biofouling is getting more serious (a shortening days), responding to the treatment (a lengthening days), or causing a change in bacterial composition (shifting RX and days values).

Cullimore's interpretation of the test results gives greater weight to days than to the RX for determining the dominant species in the sample. The rate at which the onset of a reaction occurs is considered to relate to the aggressivity of the organisms within the sample. For example, a very small population of microorganisms which are very active and can rapidly generate a reaction would be more aggressive than a large population of microorganisms which are in a passive state and do not respond so quickly. The following semi-quantitative interpretations can be made based upon the days obtained:

- <6 hours or <0.25 days — Very large (>10^7 CFU/mL) population, or extremely aggressive microbial group
- <26 hours or <1.1 days — Very significant (>10^6 CFU/mL) population, or very aggressive microbial group
- >1.1 to <4.1 days — Significant (>10^4 CFU/mL) population, or aggressive microbial group
- >4.1 to <6.1 days — >500 CFU/mL bacterial population, or potentially aggressive microbial group

• >6.1 to <14.1 days	<500 CFU/mL bacterial population, or low aggressivity.
• >14.1 days	Microbial population traumatized/population size not ascertainable or aggressivity potential very low.

Different bacterial populations can be monitored using the BART™ system:

Total Aerobic Bacteria

(TAB-BART™) indicates whether aerobic (require oxygen) bacterial activity is occurring in the water sample. One group of bacteria included in this would be the pseudomonads that often grow in consortial biofilms in association with other microbial groups. One common example is the association of pseudomonads with sulfate-reducing bacteria in corrosive biofouling. There are also *Pseudomonas* species that excrete diffusible yellow-green pigments that fluoresce in UV light. Cullimore states that if the methylene blue indicator is reduced to colorless from or "bleached" in less than 4 of incubation, then a very aggressive aerobic population is present. When reduction occurs after 5, then the aggressivity of aerobic bacteria may be considered to be very low and anaerobic activities tend to dominate. Hach, however, interprets the reactions differently than Cullimore. Hach or Culllimore did not assign RX numbers to this test, but to keep the worklists consistent, an interpretation was created for the reaction number (based upon the "to bleaching" in the Hach instruction manual).
RX 0 - no bleaching occurred after 30
RX 1 - Greater than 8 / few aerobic bacteria are present
RX 2 - 4 to 8 / aerobic bacteria are present
RX 3 - 2 to 4 / mildly aggressive aerobic bacteria
RX 4 - 1 to 2 / aggressive aerobic bacteria
RX 5 - Less than one day / very aggressive aerobic bacteria

Iron-Related Bacteria

(IRB-BART™) test detects those heterotrophic bacteria able to assimilate, accumulate or in other ways change the forms of inorganic iron. This may include reduction (anaerobic zone) and oxidation (aerobic zone) sometimes in association with biofilm formations.

RX 1, 2, 3 or 4 indicates a dominance of IRB if the days is shorter than that of SLYM or FLOR. However, if the days is longer than that of SLYM or FLOR, then heterotrophic bacteria dominate.

RX 5 is a reliable indicator of anaerobically dominated biozones and is paralleled by more aggressive (shorter days) SRB.

RX 6 or 7 indicates an enteric bacterial presence with the major genera being Enterobacter annd Klebsiella.

RX 8 or 9 may relate to aggressive pseudomonad populations.

RX 10 would indicate (particularly if days<3) a significant enteric bacterial presence. This reaction type is not on the comparator chart, but was described by Cullimore.

Sulfate-Reducing Bacteria

(SRB-BART™) test detects the sulfate-reducing bacteria by the generation of black sulfides that may form precipitates. The problems associated with these bacteria are: (1) "rotten egg" smell, which usually means that there has been a major aerobic biofouling that has removed the oxygen out of the water and allowed these bacteria to dominate. Rotten-egg smells are not created by just the SRB group but can also be sometimes generated by other bacteria such as many of the coliforms when suitable conditions occur; (2) initiates corrosive processes that causes pitting on some metallic surfaces. Because the SRB organisms cause corrosion while growing on a surface, tests on the water itself may be negative. Positive tests will occur if some of these bacteria are present in the water column due to being sloughed off the corrosion site; and (3) generation of hydrogen sulfide (H_2S) gas as metabolic product that can react with dissolved metals such as iron or manganese to generate black sulfite deposits.

When biodetectors react in less than 3, there is a very aggressive SRB population which is usually found to be associated with even more rapid reactivity in either SLYM or FLOR. If the days is 8 or more, then the population is considered to be not aggressive.

Slime-Forming Bacteria

(SLYM-BART™) are defined as those bacteria, which are able to produce extracellular polymeric substances that will produce some type of cohesive water-retaining growth mass. These bacteria do not necessarily have the ability to accumulate iron in any way other than for basic metabolic function. Cullimore states that in practical field studies, there have been many occasions where the biofouling has not been caused by IRB but by the non-iron accumulating slime forming bacteria. RX 5 is particularly common when the dominant bacteria are pseudomonads.

Fluorescing Pseudomonads

(FLOR-BART™) test specifically encourages the growth and fluorescing activity of *Pseudomonas* species. These bacteria dominate in biozones where there is available oxygen or nitrates, and a restricted range of organic compounds in relatively high concentrations (5-500 ppm). Under these conditions of organically polluted groundwaters, a restricted range of pseudomonads may dominate the biozone and trigger a very short days. Under extreme circumstances, severely polluted water well has been known to give a days of 20 minutes.

PROTOCOL USED FOR PROTOZOA ANALYSIS DURING THIS STUDY

a. Arrival. Upon arrival water samples were kept between 3 and 7°C in a walk-in cooler. Samples are examined as soon as practical, for the presence of protozoa.
b. For any protozoa that were found under the microscope, the size is measured and a general description of the protozoan is recorded, for later identification. A photograph is also taken. For this general screening, 1-2 drops from each sample were examined at 32x and 100x magnification using a phase contrast microscope manufactured by Zeiss-Jena in Germany and fitted with a Leitz/Wetzler camera. The ocular piece is 10X magnification and the objectives 3.2X, 10X, 40X, and oil immersion 100X to achieve

magnifications of 32X, 100X, 400X, and 1000X. There was a calibrated micrometer in the eyepiece to measure specimens.

c. Five mL of each sample was centrifuged in a 15 mL conical centrifuge tube in Clay Adams countertop centrifuge for 4 min at 70 rpm. 4 mL of supernatant was discarded and 1 drop of the pellet is examined in a welled slide at 32X and 100X.
d. For cultivation, modified PYNFH[2] (ATCC) culture media, was used. Modified PYNFH Medium is especially formulated for the cultivation of small, free-living amoebae, including the amoebae commonly associated with *Legionella* i.e. *Acanthamoeba*, *Hartmanella*, *Naegleria* and *Tetrahymena*. The media is provided in free-dried vials, each to be reconstituted in sterile water to make 55 mL. It was made up at double strength, i.e., one vial in 27.5 mL of sterile water. Under a biosafety hood. 2.5 mL of each water sample was added to 2.5 mL of cultivation media in a 15 mL centrifuge tube and was then incubated at 37°C. After 24 hours, 1- 2 drops were examined. After 72 hours the culture media was centrifuged and examined as in step c.
e. Samples identified as having protozoa was preserved in Bouin's fluid - in a 1:20 ratio of sample to fluid. It was then stained by first preparing the surface of the slide with albumen and letting dry slightly. One drop of culture was placed on top and quickly dried with a hair dryer. The slide was placed in 2% silver nitrate for staining and then dried by air before examination.

APPENDIX B
QUESTIONNAIRE FOR PARTICIPATING WATER UTILITIES

WELL INFORMATION CHECKLIST
PROJECT AwwaRF: "Occurrence of *Legionella* in Groundwater"

Principal Investigators:
Dr. Serge Riffard and Dr. Syed A. Sattar
Centre for Research on Environmental Microbiology
Faculty of Medicine
University of Ottawa
451 Smyth Road
Ottawa, Ontario
Canada K1H 8M5
Phone: (613) 562 5800 ext 8314
Fax: (613) 562 5452

Well Name and Location:

1. GENERAL INFORMATION

1.1 Well system owner/operator name, address:

Owner Name:

Address:

1.2 Well location (street address or general location, city, state, county)

Are there other wells at the same location?
If yes, the number of wells in the well field is:__________________

1.3 What is the size of the water supply system, in number of people served?
25 - 3,300
3,301 - 10,000
> 10,000
Unknown

1.4 Has the groundwater source for this well or well system been designated as groundwater source under the influence of surface water?

If yes, by whom (state, U.S. EPA, company, etc.)?

1.5 Is any type of treatment applied, such as disinfection, etc., before the distribution system?

If yes, type of treatment and why?

2. CONTAMINATION SOURCE CONTROLS

2.1 Contaminant Sources

2.1.1 Has the well ever been tested for microorganisms outside this project?

If yes, sample(s) collected at: Wellhead Distribution system

Date of last test:____________________

Date of last positive result:___________________

2.1.1.1 Is scheduled microbial monitoring regularly conducted on the well?

If yes, how often and for what parameters?__________________

2.1.1.2 Does this source have a historical coliform problem?

If yes, samples collected at: Wellhead Distribution system

2.1.2 Can you provide us with bacteriological analyses results on the samples collected for this study (from the well)?

Heterotrophic Plate Count?
Coliforms and total coliforms?
Others?

2.1.3 Do you conduct regular chemical analyses of your water (from the well)?

If yes, what types:

2.1.4 Will you be able to provide us with chemical analyses on the samples obtained for this study?

Relevant parameters for this study are:

Total Organic Carbon (TOC)
Dissolved Oxygen
pH
Conductivity
Turbidity
All metals usually analyzed (i.e., calcium, iron, copper, magnesium, manganese, zinc, potassium, aluminum, sodium and nickel).

2.2 Source Water Protection Program

2.2.1 Has a wellhead protection area (WHPA) been defined for the well or well field?

2.2.2 If a wellhead protection program is in place, the program status is:

Area is delineated, source inventory not completed
Area is delineated, source inventory completed
Area is delineated, source management program operating
Unknown

2.3 Inclusion of Microbial Protection

2.3.1 Are there any microbial or fecal source control programs in place, such as performance standards, setbacks, or zoning restrictions? Describe please:

2.3 Management of Potential Contaminant Sources

2.4.1 Are known contaminant sources being adequately managed

If yes, what management practices demonstrate this?

3. HYDROGEOLOGY

3.1 Groundwater Time-of-Travel

3.1.1 What aquifer is the well screened in?

3.1.2 Has the time-of-travel (TOT) of groundwater from the surface to the source aquifer been determined with confidence?

If yes, what is the TOT (years)________________

How determined?________________________

3.2 Aquifer Confinement

3.2.1 Is the source aquifer confined?

3.2.2 If confined, what is the depth from the land surface to the top of the confining layer?

Depth:________________(feet)

3.2.3 If confined, what is the minimum thickness of the confining layer?

Thickness:_____________(feet)

3.2.4 Is there a surficial aquifer or water-bearing unit above the confining layer?

If yes, what is the depth from the ground surface to the water table of the surficial aquifer?

Depth:________________(feet)

If yes, does the groundwater move upward from the source aquifer through the confining unit to the overlying aquifer, or does ground water more downward from the surficial aquifer through the underlying confining unit to the surface aquifer?

3.2.4.1 What is the diameter of the well?

3.2.5 What is the depth of the well?

3.2.6 What is the well production?

3.2.7 What is the distance to the nearest agriculture source?

3.2.9 What is the distance to the nearest river?

3.3 Aquifer Porosity

3.3.1 What is the geology of the source aquifer?

Sedimentary:	__Consolidated	__Unconsolidated
Igneous or metamorphic rock		
Volcanic:	__Consolidated	__Unconsolidated
Unknown		

3.3.2 What best describes the type of porosity of the source aquifer?

Intergranular - (e.g., sand, gravel, silt, sandstone, limestone)

Fracture/ Solution - (e.g., limestone, dolomite, granite, basalt)
Karst
Unknown
Other

3.4 Protectiveness of Vadose Zone

3.4.1 For the unconfined aquifer (or if it is unknown if the aquifer is confined), what is the depth from the land surface to the tope of water table (i.e., what is the depth to the static ground water level?)
Depth:________________(feet)

3.4.2 What best describes the lithology of the vadose zone (i.e., between the land surface and the water table?)

Gravel   sand   silt   clay   glacial tilt   alluvium
Weathered rock   sandstone   conglomerate   limestone   dolomite
Granite   basalt   metamorphic   unknown   other

3.4.3 The local average precipitation is______________(inches/year)

3.4.4 Additional known sources of recharge (other than infiltration of precipitation):\

Irrigation   wastewater application   artificial recharge
Water reclamation projects   high density septic drain fields
Other

3.5 Age of Groundwater

3.5.1 Has age dating of groundwater from the source aquifer been performed?

3.5.2 Has the groundwater travel times to the well been determined (it could be derived from the well protection program)

4. WELL CONSTRUCTION AND CONDITION

4.1 Compliance With State Regulations

4.1.1 Is a boring log and/or well construction log available for this well? If yes, please, attach a copy.

4.1.2 The bored hole diameter is:______________inches

4.1.3 What year was the well drilled?

4.1.4 Is the well screened or open-hole?

4.1.5 If screened, what is the depth of the screened interval(s)?

From_____ to_______(feet)

4.1.6 If screened, what type of filter pack was used?

4.1.7 Within the past 2 years, has the integrity of the well and distribution system been evaluated professionally?

4.1.8 Has the well been repaired or modified since installation?

4.1.9 Does the well have a backflow prevention device between the water discharge line and the distribution system?

4.2 Well Casing and Grouting

4.2.1 Is the well cased?

4.2.2 If cased, is the casing grouted in place?

4.2.3 If cased, what is the well casing material?
Concrete wood PVC low-carbon steel stainless steel
fiberglass unknown other

4.2.4 If the well is cased, is there evidence that the casing or joints are leaking (e.g., sound of water running or dripping in the well)?

4.2.5 If the well is cased, how deep does the case extend?

4.2.6 If casing is grouted, how far does the grout extend?

4.3 Existence and Condition of Sanitary Seal

4.3.1 Is there a sanitary well seal?

4.3.2 If there is a sanitary well seal, how far does it extend below the ground surface?

4.4 Surface Drainage

4.4.1 Is the well in floodplain?

If yes, what type of flood event (yearly, etc.)?

4.4.2 Is the well located in an area, which could collect surface runoff in the immediate vicinity of the well?

APPENDIX C
SAMPLE AND DATA SHEET

Sample and Data Sheet

Awwa Research Foundation Project:
Occurrence of *Legionella* in Groundwater

Our FedEx Account Number is: # ******** (Overnight shipment on ice, please. Coolers will be sent back as soon as possible.)

Please send us an email when you send the samples to sriffard@uottawa.ca so that we can have the FedEx tracking number

-Biofilm sample (Please, take care to fill the bottle(s)/or the pipe section(s) / or the water meter(s) collected containing the biofilm with raw water in order to prevent biofilm from drying).
-Water sample (4-liters) in a polypropylene container or in 4 sterile one-liter polypropylene bottles as your convenience.

Biofilm and water samples to be sent to:
Centre for Research on Environmental Microbiology
University of Ottawa, Faculty of Medicine
451, Smyth Road, Ottawa, Ontario, Canada, K1H 8M5

Contact person: Serge Riffard, Ph.D. or Syed A. Sattar, Ph.D.
Tel: 613 562 5800 ext 8314
Fax: 613 562 5452

Date:

Location (be precise please):

Sampler identity:

Please provide (if possible) the following data (field measurement):

pH during sampling:

Water temperature during sampling:

Dissolved oxygen:

Conductivity:

Turbidity:

Air temperature:

Indication(s) on local weather would be appreciated:

For those utilities that are sampling chlorinated water, please provide–if possible:

Chlorine Residual Total:

Chlorine Residual Free:

Any information related to corrosion inhibitors within the system sampled

We are very grateful for your help.
We are available for additional information.

Please don't forget to fill out Well Information CheckList for each selected well! All other relevant information is welcome. (Driller's Logs and water analyses: chemical and bacteriological analyses for the well(s) selected) Please mail them to our address (no rush!)

APPENDIX D
SAMPLE AND UTILITY DATA

Table D.1
Water and biofilm information for well #103 from Water Utility A (Phase II)

	Sample Type						
Date	Biofilm w (cm) x l (cm)	Biofilm description	Water (# of 1 L bottles)	Water description	Biofilm resuspension (mL)	Water resuspension (mL/L filtered)	Protozoa analysis
Mar-Apr (2000)	PVC pipe* 3.7 x 42.0	Black, with particulates	3	Rotten-egg odor, slight yellow color	52	10	No
Apr-May (2000)	PVC pipe† 4.8 x 36.0	Black, with particulates	3	Rotten-egg odor, slight yellow color	50	10	No
May-Jun (2000)	PVC pipe† 4.8 x 36.0	Black, with particulates	3	Rotten-egg odor, slight yellow color	100	10	No
Jun-Jul (2000)	PVC pipe† 5.1 x 41.0	Black, with particulates, rotten-egg odor	3	Rotten-egg odor, slight yellow color	50	10	Yes
Aug-Sep (2000)	PVC pipe† 5.0 x 35.0	Brown, with particulates, rotten-egg odor	3	Rotten-egg odor, clear	85	10	Yes
Nov-Dec. (2000)	PVC pipe† 5.0 x 36.0	Brown with particulates, sulfur odor	3	Rotten-egg odor, clear	150	10	Yes
Dec-Jan (2001)	PVC pipe‡ 5.0 x 38.0	Brown, particulates	3	Rotten-egg odor, slight yellow	100	10	No
	End-cap scraping	Black, particulates			100		
Jan.-Feb (2001)	PVC pipe§ 4.8 x 38.0	Brown with black sediment	3	Rotten-egg odor, clear, Black/brown particles on filter	100	10	No
	End-cap scraping	Black particles, no smell			100		
Feb-March (2001)	PVC pipe‡ 5.0 x 33.0	Light yellow with brown particles	3	Clear, slight vanilla smell	100	10	Yes
Apr-May (2001)	PVC pipe§ 5.0 x 33.0	Gray, with sediment (gray)	3	Rotten-egg odor, clear	120	10	No
May-June (2001)	PVC pipe‡ 5.0 x 36.0	Gray/black, with sediment	None	NA	180	NA	No
	Wiffle ball**	Brown, slimy strands			NA		
Jun-July (2001)	PVC pipe§ 5.0 x 31.0	Brown, rotten-egg odor	None	NA	700	NA	No
Jul-Aug (2001)	PVC pipe‡ 5.0 x 38.0	Dark brown, with sediments	3	Clear, rotten-egg odor		10	No
Aug-Sep (2001)	Wire brushing of casing	Dark brown, sulfur smell	None	NA	60	NA	No

* Biofilm development on pipe section for 7.
† Biofilm development on pipe section for 15.
‡ Biofilm development on pipe section for 1 month.
§ Biofilm development on pipe section for 2 months.

** The wiffle ball was taken from a stripping tower in the new hydrogen sulfide removal plant, only a phosphate inhibitor had been added to the raw water before this point.

Table D.2
Water and biofilm information for wells #136 and #116 from Water Utility A

Date	Sample Type				Biofilm resuspension (mL)	Water resuspension (mL/L filtered)	Protozoa analysis
	Biofilm w (cm) x l (cm)	Biofilm description	Water (# of 1 L bottles)	Water description			
Well 136							
Mar-Apr (2000)	PVC pipe* 3.7 x 38.0	Light yellow, turbid	3	Rotten-egg odor, slight yellow color	105	10	No
Apr-May (2000)	PVC pipe† 4.8 x 36.0	Turbid, odorless	3	Rotten-egg odor, slight yellow color	90	10	No
May-Jun‡ (2000)	PVC pipe† 4.8 x 36.0	Turbid, odorless	Not Received	NA	75	NA	No
Well 116							
Aug-Sep (2001)	PVC pipe§ 5.0 x 38.0	Gray, turbid, black particles	3	Clear, odorless			
Sep-Oct (2001)	PVC pipe** 5.0 x 38.0	Gray/brown, particles	3	Clear, odorless			

* Biofilm development on pipe section for 7.
† Biofilm development on pipe section for 15.
‡ 5 important storms in the 2 weeks prior to sampling after driest May in recorded history. In addition, 4 before sampling, wells were shut down and well #136 went into phase loss and could not be restarted. When this happens, the well looses pressure and water from other wells may backflow into the inactive well, possibly contaminating.
§ Biofilm development on pipe section for 1 month.
** Biofilm development on pipe section for 2 months.

Table D.3
Water and biofilm samples for well #138 from Water Utility A

	Sample Type						
Date	Biofilm w (cm) x l (cm)	Biofilm description	Water (# of 1 L bottles)	Water description	Biofilm resuspension (mL)	Water resuspension (mL/L filtered)	Protozoa analysis
Jun-Jul (2000)	PVC pipe* 4.8 x 36.0	Brown, with particles, rotten-egg odor	3	Rotten-egg odor, very turbid	50	10	Yes
Aug-Sep (2000)	PVC pipe* 5.0 x 35.0	Brown, with particulates, rotten-egg odor	3	Rotten-egg odor, clear	150	10	Yes
Nov-Dec (2000)	PVC pipe* 5.0 x 36.0	Light brown with sediment, sulfur odor	3	Rotten-egg odor, turbid	150	10	Yes
Dec-Jan (2001)	PVC pipe† 5.0 x 38.0	Brown, particulates	3	Rotten-egg odor, slight yellow	150	10	No
	End-cap scraping	Black, particulates			100		
Jan-Feb (2001)	PVC pipe‡ 5.0 x 38.0	Gray, black particles	3	Rotten-egg odor, clear	150	10	No
	End-cap scraping	Black particles, no smell		Black/brown particles on filter	100		
Feb-Mar (2001)	PVC pipe† 5.0 x 38.0	Light yellow with brown particles	3	Black sediment, slight vanilla smell	100	10	Yes
Apr-May (2001)	PVC pipe‡ 5.0 x 38.0	Orange/brown, rotten-egg odor	3	Rotten-egg odor, particles on filter	150	10	No
May-Jun (2001)	PVC pipe† 5.0 x 38.0	Brown, with particulates	3	Rotten-egg odor, clear	180	10	No
Jun-July (2001)	PVC pipe‡ 5.0 x 38.0	Turbid, gray/brown	3	Rotten-egg odor, brown sediment	700	10	No
Jul-Aug (2001)	PVC pipe† 5.0 x 38.0	Light brown, with sediments	3	Rotten-egg odor, turbid		10	No
Aug-Sep (2001)	PVC pipe† 5.0 x 38.0	Gray, black sediments	3	Clear, odorless		10	No
Sep-Oct (2001)	PVC pipe‡ 5.0 x 38.0	Gray-brown, brown particles	3	Clear, odorless		10	No

* Biofilm development on pipe section for 15.
† Biofilm development on pipe section for 1 month.
‡ Biofilm development on pipe section for 2 months.

Table D.4
Additional biofilm samples received from Water Utility A

Sampling identity/ Brief description	Biofilm width (cm) x length (cm)	Biofilm description	Biofilm resuspension (mL)	Protozoa analysis
KII RW / scraped from cacade tower in clear well before any treatment	Scraped in water	Gray/green, tough sticky strands	100	Yes
KII IFS / scraped from stand pipe after end cap of blow off valve was removed on influent main line to KII wells	End cap scraping	Black particles	100	Yes
05B / obtained from water meter in distribution system	Water Meter	Yellow/brown, black particles, odorless		No
15A / obtained from water meter in distribution system	Water Meter	Yellow/brown, brown particles		No
32G / obtained from water meter in distribution system	Water Meter	Black/gray, black particles		No
72F / obtained from water meter in distribution system	Water Meter	Gray/brown, black particles, odorless		No
94A / obtained from water meter in distribution system	Water Meter	Orange/brown, black particles, odorless		No

Table D.5
Water and biofilm information for water utilities C to Q

Well	Sample identity	Sample Type: Biofilm width (cm) x length (cm)	Biofilm description	Water (# of 1 L bottles)	Water description	Biofilm resuspension (mL)	Water resuspension (mL/L filtered)	Protozoa analysis
C	AZ(A)*	None		3	Clear, odorless	None	10	No
	AZ(B)	Water meter†	Black/green	3	Clear, odorless	115	10	
	AZ(B/M)	PVC coupons	Reddish brown	None	None	125	None	
D	EP	Water meter† (bronze)	Reddish/brown	4	Colorless, rotten-egg odor	150	10	No
E	ST*	Water meter† (brass)	Red/yellow/ brown	4	Colorless, odorless	150	10	No
F	CS2	PVC pipe‡ 2.0 x 35.5	Turbid	4	Colorless, odorless	115	10	No
	CS5	PVC pipe‡ 2.0 x 34.0	Turbid	4	Colorless, odorless	115	10	
G	PEO*	Water meter§ (brass)	Reddish, turbid	4	Colorless, odorless	200	10	No
H	PK	Water meter** (Trident/ Neptune)	Reddish, turbid	4	Clear	200	10	Yes
I	MEX*	Water meter† (Badger/Brass)	Black, turbid	4	Colorless, odorless	200	10	No
J	SF	PVC pipe†† 4.8 x 43.0	Orange/brown, odorless	4	Odorless, orange particles on filter	150	10	No
	MN	PVC pipe†† 4.8 x 43.0	Orange/brown, very thick	4	Odorless, orange particles on filter	150	10	
K	NJ*	Water meter**	Orange/brown, odorless	4	Odorless, yellow particles on filter	100	10	No
L	RI	Scraping	Orange/brown, odorless	4	Odorless, yellow particles on filter	100	10	No
M	CH*	Water meter	Odorless, black	4	Odorless, gray particles on filter	100	10	No
N	NB3	PVC pipe†† 2.0 x 35.0	Light yellow, odorless	4	Vanilla-like odor, clear	115	10	No
	NB5	PVC pipe†† 2.0 x 36.0	Clear, odorless	4	Vanilla-like odor, clear	115	10	
O‡‡	MO18	Scraping	Orange/brown (like mud)	4	Yellow, slight odor	250	10	No
	MO34	Scraping	Orange/brown (like mud)	4	Yellow, slight odor	250	10	
P	NE54	PVC pipe†† 2.3 x 30.5	Clear, odorless	4	Clear, odorless	150	10	No
	NE66	PVC pipe†† 2.3 x 30.5	Orange/brown, odorless	4	Light yellow, odorless	150	10	
Q	FK8	None	None	4	Clear, odorless	None	10	No
	FK13	None	None	4	Clear, odorless	None	10	

* Water sample from chlorinated source, contained sodium thiosulfate preservative.
† The pipe prior to the water meter was made of copper.
‡ Biofilm developed on pipe for 15.
§ The pipe prior to the water meter was made of lead.
** The pipe prior to the water meter was made of galvanized iron.
†† Biofilm developed on pipe for 1 month.
‡‡ More than 13 cm of rainfall in 3 before sampling.

APPENDIX E
FIELD PARAMETERS FOR PARTICIPATING UTILITIES

Table E.1
Field parameters for well #103 from Water Utility A

Sampling date	Pipe section biofilm growth	Water (°C)	Air (°C)	Dissolved oxygen (mg/L)	Turbidity (NTU)	pH	Conductivity (mS/m)
March-April 2000	7	24.9	27.2	0.40	0.14	7.46	48.0
April-May 2000	14	25.3	29.4	0.45	0.16	7.45	4.17
May-June 2000	19	25.0	30.0	N/D	N/D	N/D	N/D
June-July 2000	18	25.2	31.7	0.55	0.26	7.65	43.9
Aug.-Sept. 2000	18	24.9	ND	0.51	0.29	7.65	47.2
Nov.-Dec. 2000	26	ND	ND	ND	ND	ND	ND
Dec. 2000-Jan. 2001	33	24.3	23.0	0.50	0.22	7.53	45.8
Jan.-Feb. 2001	58	24.1	25.0	0.30	0.28	7.58	49.0
Feb.-March 2001	31	24.4	23.0	0.55	0.16	7.49	43.0
April-May 2001	71	24.4	23.0	0.55	0.16	7.49	43.0
May-June 2001	29	ND	ND	ND	ND	ND	ND
June-July 2001	66	ND	ND	ND	ND	ND	ND
Aug.-Sept. 2001	30	25.2	ND	0.55	0.26	7.65	43.9
Sept.-Oct. 2001	Not Applicable	ND	ND	ND	ND	ND	ND

Table E.2
Field parameters for well #136 and well #116 from Water Utility A

Sampling date	Pipe section biofilm growth	Water (°C)	Air (C)	Dissolved oxygen (mg/L)	Turbidity (NTU)	pH	Conductivity (mS/m)
Well #136							
March-April 2000	7	25.0	25.6	0.45	0.21	7.52	45.1
April-May 2000	14	25.3	28.9	0.48	0.15	7.47	42.1
May-June 2000	19	N/D	N/D	N/D	N/D	N/D	N/D
Well #116							
Aug.-Sep. 2001	30	24.9	ND	0.31	0.57	7.31	41.4
Sep.-Oct. 2001	60	25.0	ND	0.44	0.18	7.42	41.8

Table E.3
Field parameters measured for well #138 from Water Utility A

Sampling date	Pipe section biofilm growth	Water (°C)	Air (°C)	Dissolved oxygen (mg/L)	Turbidity (NTU)	pH	Conductivity (mS/m)
June-July 2000	18	25.1	31.1	0.61	0.19	7.78	43.2
August-September 2000	18	25.1	ND	0.48	0.25	7.97	45.7
November-December 2000	26	23.7	25.1	0.50	ND	7.46	ND
December-January 2001	33	24.2	22.5	ND	0.36	7.58	43.2
January-February 2001	58	24.3	25.0	0.45	0.24	7.55	47.8
February-March 2001	31	23.6	23.0	0.48	0.22	7.38	41.1
April-May 2001	71	23.6	22.5	0.48	0.22	7.38	41.1
May-June 2001	29	25.3	30	0.61	ND	7.50	ND
June-July 2001	66	25.1	31.1	0.55	0.46	7.48	38.8
July-August 2001	30	25.1	ND	0.61	0.19	7.78	43.2
August-September 2001	30	25.1	ND	0.38	0.25	7.52	43.3
September-October 2001	62	24.8	ND	0.38	0.22	7.41	41.2

Table E.4
Field parameters for water utilities involved in the study (other than Water Utility A well #103, well #116, well #136, and well #138)

Water utility	Well ID	Water (°C)	Air (°C)	Dissolved oxygen (mg/L)	Turbidity (NTU)	pH	Conductivity (mS/m)
A	KII RW	23.2	16	ND	ND	7.5	ND
	KII IFS	25.1	18	ND	ND	7.73	ND
	05B	29.4	ND	3.69	0.26	7.68	51.0
	15A	26.3	ND	3.92	0.29	7.71	50.8
	21A	26.3	ND	2.03	0.33	7.65	51.0
	32G	26.6	ND	5.82	0.33	7.78	51.9
	72F	26.3	ND	3.99	0.24	7.67	52.5
	94A	26.3	ND	6.43	0.30	7.84	50.4
C	AZ	33	28	ND	ND	7.7	ND
D	EP	25	ND	ND	0.19	8.2	94.6
E	ST	20	22.2	5.3	ND	7.2	ND
F	CS2	44.6	ND	0.86	0.09	8.12	ND
	CS5	40.5	ND	0.55	0.08	7.82	ND
G	PEO	ND	ND	ND	ND	ND	ND
H	PK	13.3	ND	ND	ND	7.2	ND
I	MEX	14.4	ND	ND	0.39	7.88	0.688
J	MN	11.8	-6.7	1.5	4.0	7.4	80.4
	SF	10.9	-6.7	5.8	2.3	7.3	99.4
K	NJ	12.3	ND	ND	ND	7.70	ND
L	RI	9.0	ND	ND	0.15	7.6	ND
M	CH	15.0	ND	ND	ND	ND	ND
N	NB3	ND	ND	ND	ND	ND	ND
	NB7	ND	ND	ND	ND	ND	ND
O	MO18	14.5	23.4	4.1	21.9	7.47	64.3
	MO34	15.7	23.4	1.9	35.2	7.10	67.9
P	NE54	14.4	31.1	ND	0.14	7.2	45.2
	NE66	15.0	31.1	ND	0.28	7.3	59.0
Q	FK 8	24.7	28	3.59	0.28	7.4	59.7
	FK 13	24.0	28	4.79	0.31	7.4	63.0

APPENDIX F
CLUSTERING FOR STRAINS ISOLATED IN PHASE I

Identification by phylogenetic positioning of some *Legionella* isolates on the basis of their 16S rRNA gene sequences (fragment size of 531-bp) by comparison with available sequences of described species.

Isolates from groundwater samples are indicated in bold.

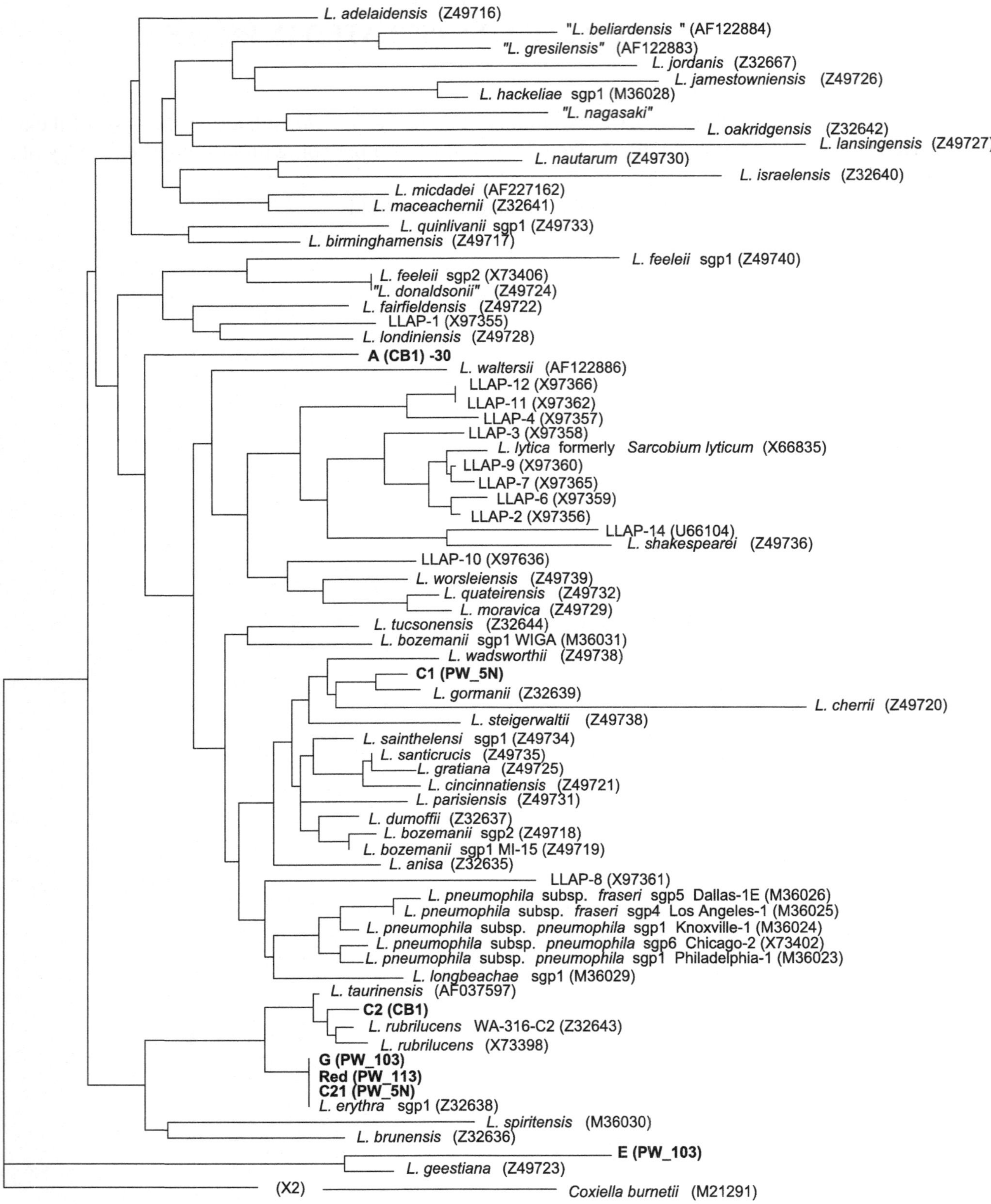

L. adelaidensis (Z49716)
"L. beliardensis " (AF122884)
"L. gresilensis" (AF122883)
L. jordanis (Z32667)
L. jamestowniensis (Z49726)
L. hackeliae sgp1 (M36028)
"L. nagasaki"
L. oakridgensis (Z32642)
L. lansingensis (Z49727)
L. nautarum (Z49730)
L. israelensis (Z32640)
L. micdadei (AF227162)
L. maceachernii (Z32641)
L. quinlivanii sgp1 (Z49733)
L. birminghamensis (Z49717)
L. feeleii sgp1 (Z49740)
L. feeleii sgp2 (X73406)
"L. donaldsonii" (Z49724)
L. fairfieldensis (Z49722)
LLAP-1 (X97355)
L. londiniensis (Z49728)
A (CB1) -30
L. waltersii (AF122886)
LLAP-12 (X97366)
LLAP-11 (X97362)
LLAP-4 (X97357)
LLAP-3 (X97358)
L. lytica formerly Sarcobium lyticum (X66835)
LLAP-9 (X97360)
LLAP-7 (X97365)
LLAP-6 (X97359)
LLAP-2 (X97356)
LLAP-14 (U66104)
L. shakespearei (Z49736)
LLAP-10 (X97636)
L. worsleiensis (Z49739)
L. quateirensis (Z49732)
L. moravica (Z49729)
L. tucsonensis (Z32644)
L. bozemanii sgp1 WIGA (M36031)
L. wadsworthii (Z49738)
C1 (PW_5N)
L. gormanii (Z32639)
L. cherrii (Z49720)
L. steigerwaltii (Z49738)
L. sainthelensi sgp1 (Z49734)
L. santicrucis (Z49735)
L. gratiana (Z49725)
L. cincinnatiensis (Z49721)
L. parisiensis (Z49731)
L. dumoffii (Z32637)
L. bozemanii sgp2 (Z49718)
L. bozemanii sgp1 MI-15 (Z49719)
L. anisa (Z32635)
LLAP-8 (X97361)
L. pneumophila subsp. fraseri sgp5 Dallas-1E (M36026)
L. pneumophila subsp. fraseri sgp4 Los Angeles-1 (M36025)
L. pneumophila subsp. pneumophila sgp1 Knoxville-1 (M36024)
L. pneumophila subsp. pneumophila sgp6 Chicago-2 (X73402)
L. pneumophila subsp. pneumophila sgp1 Philadelphia-1 (M36023)
L. longbeachae sgp1 (M36029)
L. taurinensis (AF037597)
C2 (CB1)
L. rubrilucens WA-316-C2 (Z32643)
L. rubrilucens (X73398)
G (PW_103)
Red (PW_113)
C21 (PW_5N)
L. erythra sgp1 (Z32638)
L. spiritensis (M36030)
L. brunensis (Z32636)
E (PW_103)
L. geestiana (Z49723)
(X2)
Coxiella burnetii (M21291)

APPENDIX G
CLUSTERING FOR STRAINS ISOLATED DURING PHASE II

Phylogenetic positioning of some *Legionella* isolates based on their 16S rRNA gene sequences (fragment size of 531-bp) by comparison with available sequences of described species. Isolates from groundwater samples during PHASE I are in bold. Some isolates obtained during Phase II are underlined.

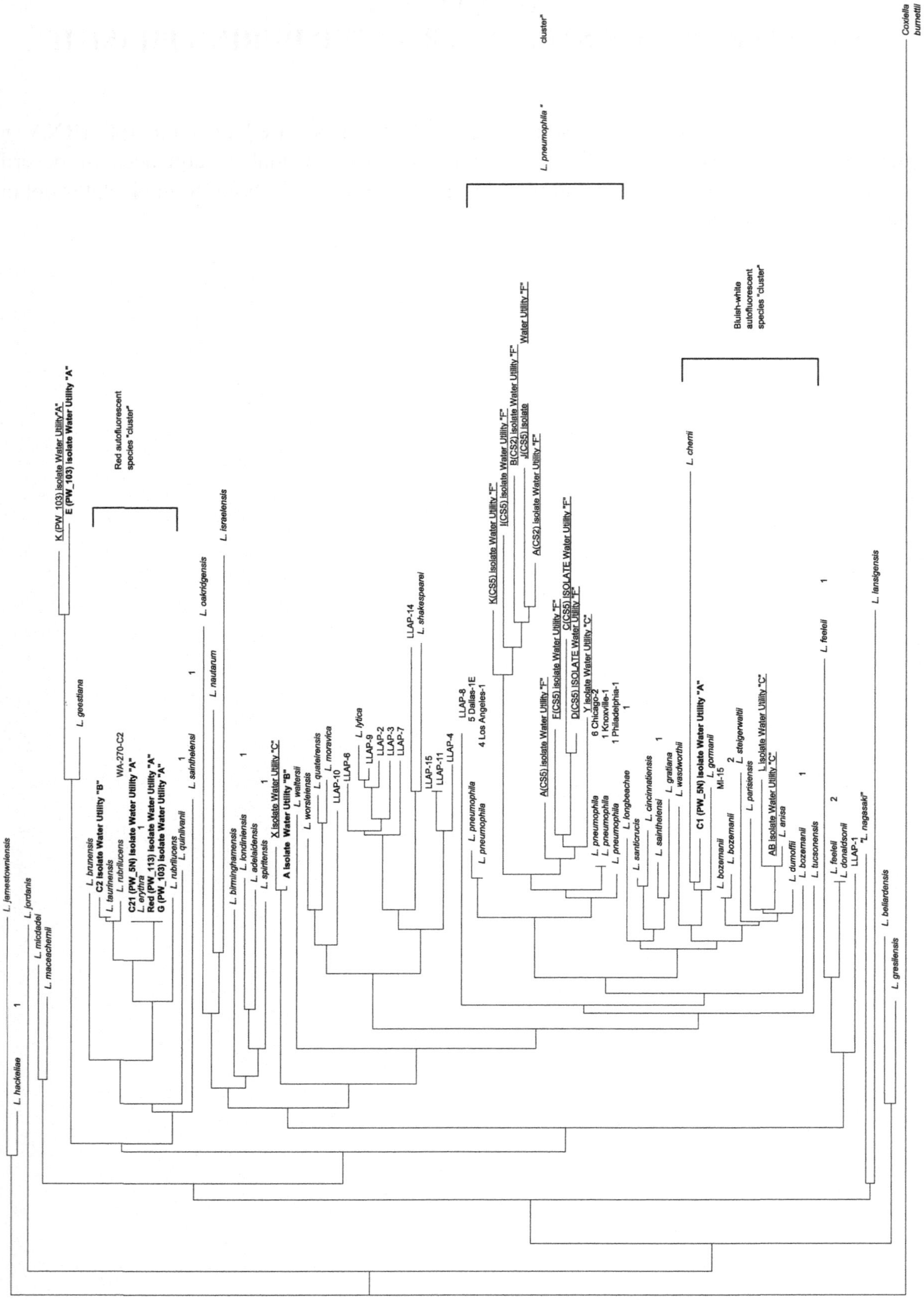
L. jamestowniensis
1
L. hackeliae
L. jordanis
L. micdadei
L. maceachernii
L. brunensis
C2 Isolate Water Utility "B"
L. tauriensis
L. rubrilucens
C21 (PW_5N) Isolate Water Utility "A"
L. erythra
1
Red (PW_113) Isolate Water Utility "A"
G (PW_103) Isolate Water Utility "A"
L. rubrilucens
L. quinlivanii
WA-270-C2
Red autofluorescent species "cluster"
K (PW_103) isolate Water Utility "A"
E (PW_103) isolate Water Utility "A"
L. geestiana
1
L. sainthelensi
1
L. oakridgensis
L. nautarum
L. israelensis
L. birminghamensis
L. londiniensis
L. adelaidensis
L. spiritensis
1
X isolate Water Utility "C"
A isolate Water Utility "B"
L. waltersii
1
L. worsleiensis
L. quateirensis
L. moravica
LLAP-10
LLAP-6
L. lytica
LLAP-9
LLAP-2
LLAP-3
LLAP-7
LLAP-14
L. shakespearei
LLAP-15
LLAP-11
LLAP-4
L. pneumophila
L. pneumophila
LLAP-8
5 Dallas-1E
4 Los Angeles-1
K(CS5) isolate Water Utility "F"
I(CS5) isolate Water Utility "F"
B(CS2) isolate Water Utility "F"
J(CS5) isolate Water Utility "F"
A(CS2) isolate Water Utility "F"
L. pneumophila " cluster"
A(CS5) isolate Water Utility "F"
F(CS5) isolate Water Utility "F"
D(CS5) ISOLATE Water Utility "F"
C(CS5) ISOLATE Water Utility "F"
Y isolate Water Utility "C"
6 Chicago-2
1 Knoxville-1
1 Philadelphia-1
1
L. pneumophila
L. pneumophila
L. pneumophila
L. longbeachae
L. santicrucis
L. cincinnatiensis
L. sainthelensi
L. gratiana
1
L. wadsworthii
L. cherrii
C1 (PW_5N) Isolate Water Utility "A"
L. gormanii
MI-15
L. bozemanii
L. bozemanii
2
L. steigerwaltii
L. parisiensis
L isolate Water Utility "C"
AB isolate Water Utility "C"
L. anisa
1
L. dumoffii
L. bozemanii
L. tucsonensis
Bluish-white autofluorescent species "cluster"
L. feeleii
1
L. feeleii
2
L. donaldsonii
LLAP-1
"L. nagasakii"
L. lansingensis
L. beliardensis
L. gresilensis
Coxiella burnetii

APPENDIX H
RESULTS OF PROTOZOA ANALYSIS

Table H.1
Protozoa analysis for water utility "A" from well PW_103 for samples received between June-July and November-December 2000

	Treatment			
Sample	Fresh water	Fresh sample, centrifuged pellet	Culture in PYNFH medium	Culture in rice
PW_103 Biofilm (June-July)	Fibrous plant life (not spiny), free living worm	Fibrous plant life (not spiny)	Lots of black debris	Algal cells, lots of black debris
PW_103 Water (June-July)	Algal cells, fibrous plant life	Nothing noted	Nothing noted	Algal cells
PW_103 Biofilm (Aug-Sept)	Nothing noted		Plant material	Nothing noted
PW_103 Water (Aug-Sept)	Plant material		Ciliate (*Colpoda*)*	Nothing noted
PW_103 Biofilm (Nov-Dec)	Plant material		Plant material	Nothing noted
PW_103 Water (Nov-Dec)	Some plant material		Plant material	Nothing noted

* *Colpoda* have not been previously reported to harbour legionellae. This isolation does not, therefore, necessarily indicate any such association.

Table H.2
Protozoa analysis for water utility "A" from well PW_103 for samples received between December 2000 and March 2001

	Treatment			
Sample	Fresh sample or fresh centrifuged pellet	Incubated at 23.5°C for 48-72 hours	Incubated at 23.5°C for an additional 72 hours (5-6 total)	Incubated at 20°C for an additional 72 hours (8-9 total)
PW_103 Biofilm pipe	Sand particles	Flagellates (6-10 μm)	Flagellates	Flagellates
Biofilm end cap (Dec-Jan)	Black debris	Black debris	Nothing	
PW_103 Water (Dec-Jan)	Nothing	Nothing	Nothing	
PW_103 Biofilm pipe	Broke in transit	Broke in transit	Broke in transit	Broke in transit
Biofilm end cap (Jan-Feb)	Black debris, plant material	Black debris, plant material		
PW_103 Water (Jan-Feb)	Nothing	Nothing	Nothing	
PW_103 Biofilm pipe	Black debris, plant material	*Naegleria* trophozoites, flagellates	Flagellates, no amoebic trophozoites	
Biofilm end cap (Feb-March)	ND	ND	ND	
PW_103 Water (Feb-March)	Plant material		Flagellates	

Table H.3
Protozoa analysis for water utility "A" from well PW_138 for samples received between June-July and November-December 2000

	Treatment			
Sample	Fresh water	Fresh sample, centrifuged pellet	Culture in PYNFH medium	Culture in rice
PW_138 Water (June-July)	Algal cells, fibrous spiny plant life	Algal cells, fibrous spiny plant life	Algal cells	Rotifers, worm ciliates (maybe *Tetrahymena*) or more likely *Microthorax*, amoebic trophozoites (unidentified)
PW_138 Biofilm (June-July)	No algal cells, fibrous spiny plant life	Plant life	Algal cells, plant life	Algal cells
PW_138 Water (August-September)	Plant material		Plant material	Plant material
PW_138 Biofilm (August-September)	Nothing noted		Plant material	Nothing noted
PW_138 Water (November-December)	Plant material		Plant material	Nothing noted
PW_138 Biofilm (November-December)	Plant material, black debris		Plant material, black debris	Nothing noted

Table H.4
Protozoa analysis for water utility "A" from well PW_138 for samples received between December 2000 and March 2001

	Treatment			
Sample	Fresh sample or fresh centrifuged pellet	Incubated at 23.5°C for 48-72 hours	Incubated at 23.5°C for an additional 72 hours (5-6 total)	Incubated at 20°C for an additional 72 hours (8-9 total)
PW_138 Water (Dec-Jan)	Nothing	Nothing	Nothing	
PW_138 Biofilm pipe Biofilm end cap (Dec-Jan)	Nothing Black debris	Nothing Black debris, plant material	Nothing *Hartmannella* trophozoites	*Hartmannella* trophozoites and cysts
PW_138 Water (Jan-Feb)	Black debris, plant material	Black debris, plant material		
PW_138 Biofilm pipe Biofilm end cap (Jan-Feb)	Plant material Black debris, plant material	Plant material Black debris, plant material		
PW_138 Water (Feb-March)	Plant material	*Colpoda* and small ciliates, plant material	*Colpoda* and small ciliates, plant material	
PW_138 Biofilm pipe Biofilm end cap (Feb-March)	Black debris	*Naegleria* trophozoites, black debris	*Colpoda*, amoebic cysts but no amoebic trophozoites	

Table H.5
Protozoa analysis for water utility "F" samples

	Treatment			
Sample	Fresh water	Fresh sample, centrifuged pellet	Culture in PYNFH medium	Culture in rice
Well #5 Water 26/07	No algal cells or plant life	Nothing noted	Nothing noted	Not done
Well #5 Biofilm 26/07	Algal cells, plant life	Plant life	Algal cells	Not done
Well #2 Water 26/07	Nothing noted	Nothing noted	Nothing noted	Not done
Well #2 Biofilm 26/07	Algal cells, plant life	Plant life, free living nematode	Algal cells more numerous than other samples	Not done

Table H.6
Protozoa analysis for water utility "H" samples

	Treatment			
Sample	Fresh water	Fresh sample, centrifuged pellet	Culture in PYNFH medium	Culture in rice
Biofilm	Nothing noted	Plant material, a few amoebic cysts	Nothing noted	Not done
Water	Nothing noted	Little plant life, amoebic cysts	Nothing noted	Not done

APPENDIX I
ADDITIONAL STATISTICAL ANALYSES

FACTORIAL ANALYSIS (SCHEMATIC DIAGRAMS)

Schematic representations of the factorial analysis generated by the scatter module from ADE 4.0 are provided in Figure I.1 and Figure I.2. The analysis represents two separate data sets. ADE 4.0 computes the gravity center of each group of points and draws lines connecting each element to its gravity center. For each point on the factor map, the software draws a circle or a square with a size proportional to a series of values (squares represent all the wells investigated - and the values for each parameters investigated -; squares are related or not to circles; circle number 1 is the "*Legionella* presence" criterion and circle number 2 is the "*Legionella* absence" criterion).

Gravity centers (circles #1 and #2) are very close from each other on both schematic representation suggesting that the data collected cannot differentiate between contaminated wells from non contaminated ones despite the presence of additional observations for the second set of data selected.

Analysis of Indicators in Combination

Selected parameters were investigated in 2-fold combination to determine their influence, if any, on the presence of *Legionella* in the selected groundwater wells (data not shown). No such pairwise influences were found.

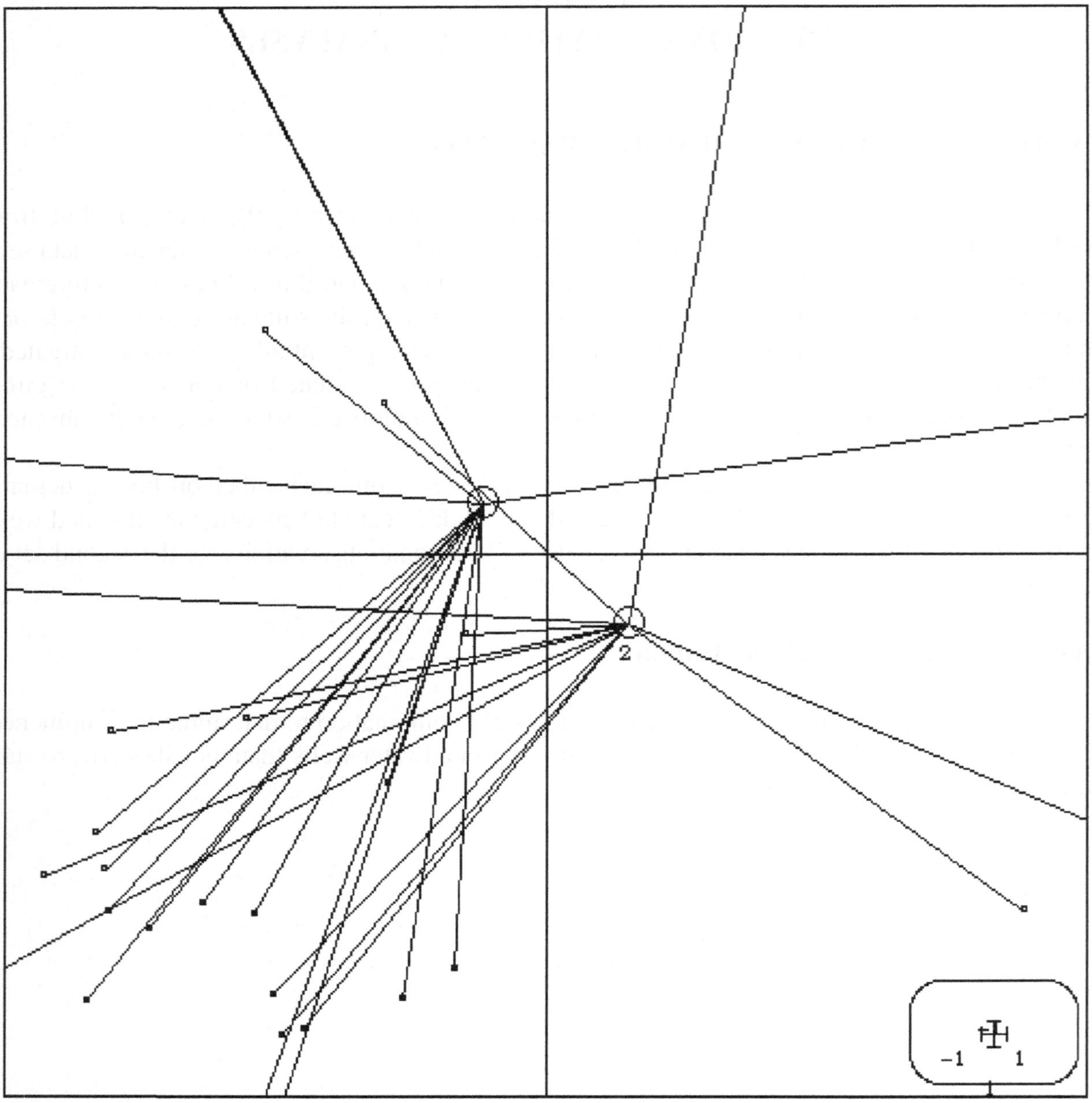

Figure I.1 Schematic representation for the first set of data (list of 14 variables with 30 observations) generated by factorial discriminative analysis

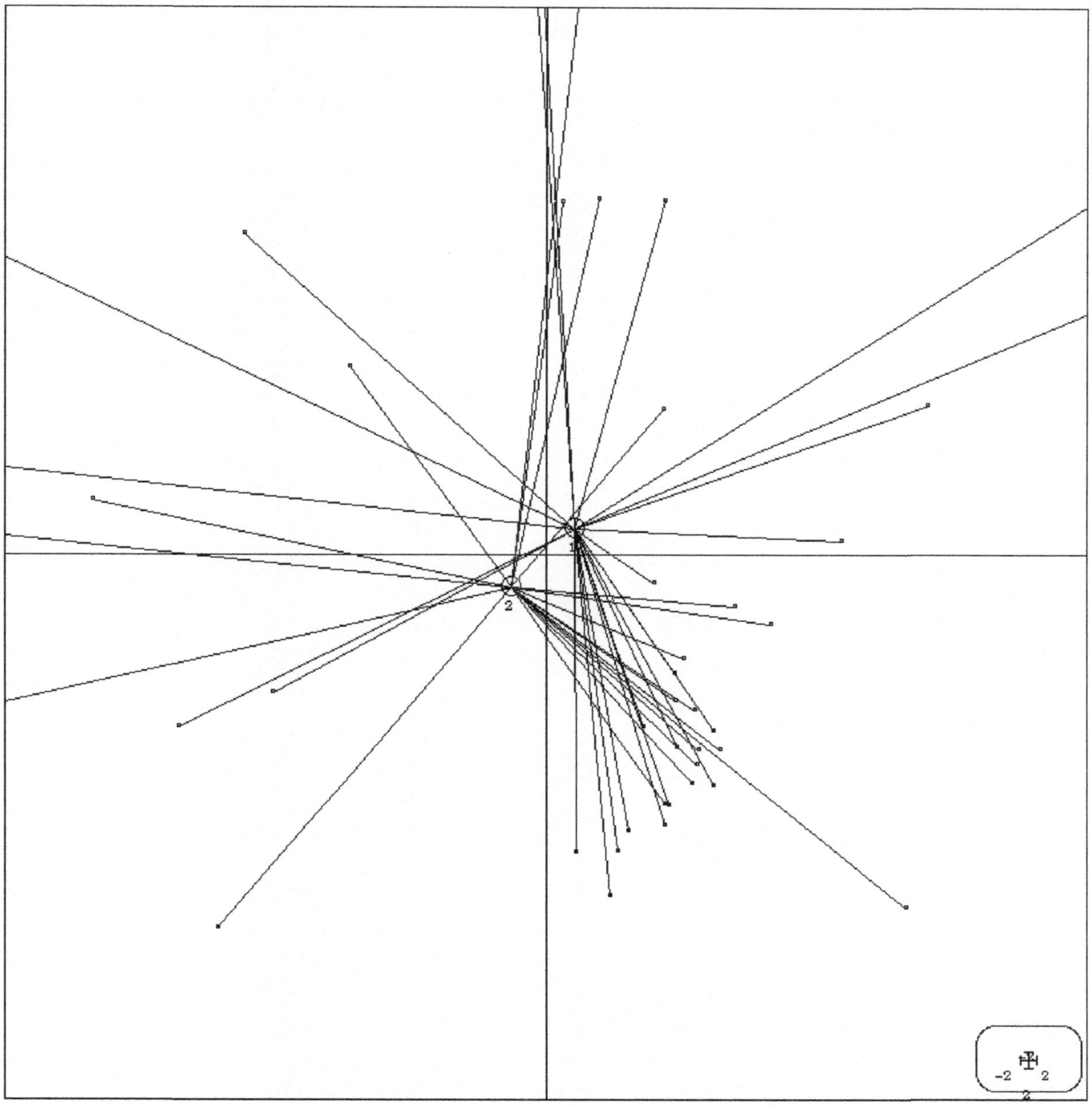

Figure I.2 Schematic representation for the second set of data (list of 5 variables with 44 observations) generated by factorial discriminative analysis

APPENDIX J
OUTREACH REPORT

Presentations and publications related to this report:

- Parts of the results obtained during Phase I in conjunction with some early work on IMS were presented orally at the Ontario Water Works Association and Ontario Municipal Water Association Joint Annual Conference (May 7-10, 2000, Windsor, Ontario). This also culminated in a student report to the Canadian Infectious Diseases Society: Douglas, S, Brooks, T., Riffard, S., Springthorpe, S., Sattar, S.A. (2000). Evaluation of immunomagnetic separation for the specific recovery of *Legionella* in samples of drinking water and biofilms.
- Parts of the results obtained were presented as a poster at the 1st World Congress of the International Water Association (Health-related Water Microbiology Symposium) in Paris, in July 2000. Riffard, S., Douglas, S., Springthorpe, S., Filion, L.G., and Sattar, S. A. (2000). Occurrence of Legionella in groundwater: an ecological study.
- Part of the results obtained since the beginning of Phase II has been presented (oral presentation) at the 16th Eastern Canadian Symposium on Water Pollution Research held in Ottawa, November 17, 2000. Brooks, T., Riffard, S., Douglas, S., Springthorpe, S., Filion, L.G. and Sattar, S.A. (2000). Occurrence of Legionella in groundwater: an ecological study.
- Parts of the results have been presented to the Microbial/Disinfection By-Products Symposium, Lisle, Illinois, March 25, 2001. Pryor, M., Riffard, S., Springthorpe, S., Sattar, S.A., and Powell, RM (2001). A microbiological assessment of groundwater and a chlorinated distribution system by a participating utility.
- Part of the results has been presented to the Ontario Water Works Association and Ontario Municipal Water Association Joint Annual Conference, Toronto, May 6-9, 2001. Brooks, T., Riffard, S., Douglas, S., Springthorpe, S., Filion, L.G., and Sattar, S.A. (2001). Isolation and characterization of *Legionella* from groundwater.
- Results of this study have been presented orally at the 10th National Drinking Water Conference, Halifax, April 2002. Brooks, T., Riffard, S., Osicki, R., Springthorpe, S. and Sattar, S. A. (2002). Occurrence of *Legionella* species in groundwater samples.
- Results of the study have been presented orally at the International Conference on Water and Health, Drinking Water Safety: A Total Quality Management Approach, Ottawa Sept 23-25, 2002. A manuscript based on this presentation will appear in a Special Issue of the Journal of Toxicology and Environmental Health as well as a NERAM (Network for Environmental Risk Assessment and Management) document. T. Brooks, R. Osicki, V.S. Springthorpe, S.A. Sattar, L. Filion, D. Abrial and S. Riffard (2002). Detection and identification of *Legionella* species from groundwaters.
- A manuscript based on this study is being prepared for submission to Applied and Environmental Microbiology.

- A portion of this work forms a part of a presentation to be made at the Specialist Group on Health Related Environmental Microbiology of the IWA to be held in Capetown in September 2003. A manuscript will also be submitted for publication following this meeting in Proceedings in Water Science and Technology. M. Pryor, S. Springthorpe, S. Riffard, T. Brooks, Y. Huo, G. Davis, S.A. Sattar (2003). Investigation of opportunistic pathogens in municipal drinking water under different supply and treatment regimes.
- A portion of this work is part of an abstract submitted to the ASM Biofilm Conference to be held Nov 1-6, 2003, in Victoria, BC.

REFERENCES

Abbaszadegan, M., M. LeChevallier, J. Rosen, L. Zimmerman, and K. Merrell. 1998. Microbial and chemical measurements for ground water vulnerability assessment. *Proceedings of the Water Quality Technology Conference*, Nov 1-4, 1998. San Diego, CA: American Water Works Association.

Abu Kwaik, Y., L.-Y. Gao, B.J. Stone, C. Venkataraman, and O.S. Harb. 1998. Invasion of protozoa by *Legionella pneumophila* and its role in bacterial ecology and pathogenesis. *Appl. Environ. Microbiol.* 64:3127-3133.

Adeleke, A. 1996. *Legionella*-like amoebal pathogens—Phylogenetic status and possible role in respiratory disease. *Emerging Infect. Dis.* 2(3): July-September.

Adeleke, A. A., B. S. Fields, R. F. Benson, M. I. Daneshvar, J. M. Pruckler, R. M. Ratcliff, T. G. Harrison, R. S. Weyant, R. J. Birtles, D. Raoult, and M. A. Halablab. 2001. *Legionella drozanskii* sp. nov., *Legionella rowbothamii* sp. nov. and *Legionella fallonii* sp. nov.: Three unusual new *Legionella* species. *Int. J. Syst. Evol. Microbiol.* 51:1151-1160.

Altschul, S. F., W. Gish, W. Miller, E.W. Myers, and D.J. Lipman. 1990. Basic local alignment search tool. *J. Mol. Biol.* 215:403-410.

Atherholt, T. B., M. W. LeChevallier, W. O. Norton and J. S. Rosen. 1998. Effect of rainfall on *Giardia* and *Cryptosporidium*. *Jour. AWWA* 90:66-80.

Atlas, R. M., J. F. William, and M.K. Huntington. 1995. *Legionella* contamination of dental-unit waters. *Appl. Environ. Microbiol.* 61:1208-1213.

Bangsborg, J. M., S. Uldum, J.S. Jensen, and B.G. Bruun. 1995. Nosocomial legionellosis in three heart-lung transplant patients: Case reports and environmental observations. *Eur. J. Clin. Microbiol. Infect. Dis.* 14:99-104.

Barbaree, J. M., B. S. Fields, J.C. Feeley, G.W. Gorman, and W.T. Martin. 1986. Isolation of protozoa from water associated with a Legionellosis outbreak and demonstration of intracellular multiplication of *Legionella*. *Appl. Environ. Microbiol.* 51:422-424.

Barer, M. R., L. T. Gribbon, C. R. Harwood, and C. E. Nwogun. 1993. The viable but non-cultivable hypothesis and medical bacteriology. *Rev. Med. Microbiol.* 4:183-191.

Bartlett, C. L. R., A. D. Macrae, and J.T. Macfarlane. 1986. Clinical aspects and diagnosis of *Legionella* infections. In *Legionella Infections*. Edited by C. L. R. Bartlett, A. D. Macrae, and J. T. Macfarlane. London: Edward Arnold. 37-55.

Baty, V., B. Hoen, H. Schuhmacher, C. Amiel, M. Reyrolle, H. Garin, and P. Canton. 1997. *Legionella jordanis* pneumonia unresponsive to fluoroquinolones in a non-immunocompromised host. *Scand. J. Infect. Dis.* 29:319-320.

Bej, A. K., M. H. Mahbubani, and R. M. Atlas. 1991. Detection of viable *Legionella pneumophila* in water by polymerase chain reaction and gene probe methods. *Appl. Environ. Microbiol.* 57:597-600.

Bej, A. K., M. H. Mahbubani, R. Miller, J.L. DiCesare, L. Haff, and R. M. Atlas. 1990. Multiplex PCR amplification and immobilized capture probes for detection of bacterial pathogens and indicators in water. *Mol. Cell. Probes* 4:353-365.

Benson, R. F., W. J. Drozanski,T. J. Rowbotham, I. Bialkowska, D. Losos, and J. C. Butler. 1995. Serologic evidence of infection with 9 *Legionella*-like amoebal pathogens in pneumonia patients. Presented at 95th Annual General Meeting of the American Society for Microbiology, Washington, DC, USA: American Society for Microbiology.

Benson, R. F., W. L. Thacker, M.I. Daneshvar, and D.J. Brenner. 1996. *Legionella waltersii* sp. nov. and an unnamed *Legionella* genomospecies isolated from water in Australia. *Int. J. Syst. Bacteriol.* 46:631-634.

Benson, R. F., W. L. Thacker, F. C. Fang, B. Kanter, W. R. Mayberry, and D. J. Brenner. 1990. *Legionella sainthelensi* serogroup 2 isolated from patients with pneumonia. *Res. Microbiol.* 141:453-463.

Benson, R. F., W. L. Thacker, J. A. Lanser, N. Sangster, W. R. Mayberry, and D. J. Brenner. 1991. *Legionella adelaidensis*, a new species isolated from cooling tower water. *J. Clin. Microbiol.* 29:1004-1006.

Benson, R. F., W. L. Thacker, R. P. Waters, P.A. Quinlivan, W. R. Mayberry, P. J. Brenner, and H. W. Wilkinson. 1989. *Legionella quinlivanii* sp. nov. isolated from water. *Curr. Microbiol.* 18:195-197.

Benson, R. F., W. L. Thacker, H. W. Wilkinson, R. J. Fallon, and D. J. Brenner. 1988. *Legionella pneumophila* serogroup 14 isolated from patients with fatal pneumonia. *J. Clin. Microbiol.* 26:382.

Bentham, R. H. 2000. Routine sampling and the control of *Legionella* spp. in cooling tower water systems. *Curr. Microbiol.* 41:271-275.

Bentham, R. H., and C. R. Broadbent. 1993. A model for automn outbreaks of legionnaires' disease associated with cooling towers, linked to system operation and size. *Epidemiol. Infect.* 111:287-295.

Bercovier, H., A. G. Steigerwalt, M. Oerhi-Cochin, C. W. Moss, H. W. Wilkinson, R. F. Benson, and D. J. Brenner. 1986. Isolation of legionellae from oxidation ponds and fishponds in Israel and description of *Legionella israelensis* sp. nov. *Int. J. Syst. Bacteriol.* 36:368-371.

Berendt, R. F. 1981. Influence of blue-green algae (cyanobacteria) on survival of *Legionella pneumophila* in aerosols. *Infect. Immun.* 32:690-692.

Berendt, R. F., H. W. Young, R. G. Allen, and G. L. Knutsen. 1980. Dose-response of guinea pigs experimentally infected with aerosols of *Legionella pneumophila*. *J. Infect. Dis.* 141:186-192.

Berk, S. G., R. S. Ting, G.W. Turner, and R.J. Ashburn. 1998. Production of respirable vesicles containing live *Legionella pneumophila* cells by two *Acanthamoeba* spp. *Appl. Environ. Microbiol.* 64:279-286.

Bhopal, R. 1995. Source of infection for sporadic legionnaires' disease: A review. *J. Infect.* 30:9-12.

Bhopal, R. S., and G. Barr. 1990. Maintenance of cooling towers following two outbreaks of Legionnaires' disease in a city. *Epidemiol. Infect.* 104:29-38.

Bibb, W. F., P. M. Arnow, D. L. Dellinger, and S. R. Perryman. 1983. Isolation and characterization of a seventh serogroup of *Legionella pneumophila*. *J. Clin. Microbiol.* 17:346-348.

Bibb, W. F., R. J. Sorg, B. M. Thomason, M. D. Hicklin, A. G. Steigerwalt, D. J. Brenner, and M. R. Wulf. 1981. Recognition of a second serogroup of *Legionella longbeachae*. *J. Clin. Microbiol.* 14:674-677.

Birtles, R. J., N. Doshi, N. A. Saunders, and T. G. Harrison. 1991. Second serogroup of *Legionella quinlivanii* isolated from two unrelated sources in the United Kingdom. *J. Appl. Bacteriol.* 71:402-406.

Bissett, M. L., J. O. Lee, and D. S. Lindquist. 1983. New serogroup of *Legionella pneumophila*, serogroup 8. *J. Clin. Microbiol.* 17:887-891.

Bitton, G., and C. P. Gerba. 1984. Groundwater pollution microbiology. In *Groundwater Pollution Microbiology*. Edited by G. Bitton and C. P. Gerba. New York: John Wiley & Sons. 1-7.

Bohach, G. A., and I. S. Snyder. 1983a. Characterization of surfaces involved in adherence of *Legionella pneumophila* to *Fischerella* species. *Infect. Immun.* 42:318-325.

Bohach, G. A., and I. S. Snyder. 1983b. Cyanobacterial stimulation of growth and oxygen uptake by *Legionella pneumophila*. *Appl. Environ. Microbiol.* 46:528-531.

Bollin, G. E., J. F. Plouffe, M. F. Para, and B. Hackman. 1985. Aerosols containing *Legionella pneumophila* generated by shower heads and hot-water faucets. *Appl. Environ. Microbiol.* 50:1128-1131.

Bopp, C. A., J. W. Sumner, G.K Morris, and J.G. Wells. 1981. Isolation of *Legionella* spp. from environmental water samples by low-pH treatment and use of a selective medium. *J. Clin. Microbiol.* 13:714-719.

Bornstein, N., and J. Fleurette. 1992. Les *Legionella*. *Lyon Pharmaceutique* 43(2):101-116.

Bornstein, N., D. Marmet, and M. Surgot. 1988. Risque de Légionelloses chez les curistes fréquentant un établissement thermal. *La lettre de l'infectiologue* 3:435-438.

Bornstein, N., D. Marmet, M. Surgot, M. Nowicki, H. Meugnier, J. Fleurette, E. Ageron, F. Grimont, P.A. Grimont, and W.L. Thacker. 1989. *Legionella gratiana* sp. nov. isolated from French spa water. *Res. Microbiol.* 140:541-552.

Bornstein, N., C. Vieilly, M. Nowicki, J. C. Paucod, and J. Fleurette. 1986. Epidemiological evidence of Legionellosis transmission through domestic hot water supply systems and possibilities of control. *Isr. J. Med. Sciences* 22:655-661.

Boulanger, C. A., and P. H. Edelstein. 1995. Precision and accuracy of recovery of *Legionella pneumophila* from seeded tap water by filtration and centrifugation. *Appl. Environ. Microbiol.* 61:1805-1809.

Breiman, R. F. 1993. Modes of transmission in epidemic and nonepidemic *Legionella* infection: Direction for further study. In *Legionella: Current Status and Emerging Perspectives*. Edited by J. M. Barbaree, R. F. Breiman, and A. P. Dufour. Washington D.C.: American Society for Microbiology. 30-35.

Brenner, D. J. 1986. Classification of Legionellaceae. Current status and remaining questions. *Isr. J. Med. Sci.* 22:620-632.

Brenner, D. J., J. C. Feeley and R. E. Weaver. 1984. Family VII. Legionellaceae. In *Bergey's Manual of Systematic Bacteriology.* Edited by N. R. Kreig and J. G. Holt. Baltimore: The Williams and Wilkins Co. 1:279-288.

Brenner, D. J., A. G. Steigerwalt, P. Epple, W. F. Bibb, R. M. McKinney, R. W. Starnes, J. M. Colville, R. K. Selander, P. H. Edelstein, and C. W. Moss. 1988. *Legionella pneumophila* serogroup Lansing 3 isolated from a patient with fatal pneumonia and descriptions of *L. pneumophila* subsp. *L. pneumophila* subsp. nov., *L. pneumophila* subsp. *fraseri* subsp. nov., and *L. pneumophila* subsp. *pascullei* subsp. nov. *J. Clin. Microbiol.* 26:1695-1703.

Brenner, D. J., A. G. Steigerwalt, G. W. Gorman, R. E. Weaver, J. C. Feeley, L. G. Cordes, H. W. Wilkinson, C. Patton, B. M. Thomason, and K. R. Lewallen Sasseville. 1980. *Legionella bozemanii* sp. nov. and *Legionella dumoffii* sp. nov.: Classification of two additional species of *Legionella* associated with human pneumonia. *Curr. Microbiol.* 4:111-116.

Brenner, D. J., A. G. Steigerwalt, G. W. Gorman, H. W. Wilkinson, W. F. Bibb, M. Hackel, R. L. Tyndall, J. Campbell, J. C. Feeley, W. L. Thacker, P. Skaliy, W. T. Martin, B. J. Brake, B. S. Fields, H. U. McEachern, and L. K. Corcoran. 1985. Ten new species of *Legionella*. *Int. J. Syst. Bacteriol.* 35:50-59.

Brenner, D. J., A. G. Steigerwalt, and J.E. McDade. 1979. Classification of the Legionnaires' disease bacterium: *Legionella pneumophila,* genus novum, species nova, of the family Legionellaceae, family nova. *Ann. Intern. Med.* 90:656-658.

Brieland, J., M. McClain, L. Heath, C. Chrisp, G. Huffnagle, M. LeGendre, M. Hurley, J. Fantone, and C. Engleberg. 1996. Coinoculation with *Hartmannella vermiformis* enhances replicative *Legionella pneumophila* lung infection in a murine model of legionnaires' disease. *Infect. Immun.* 64:2449-2456.

Brieland, J. K., J. C. Fantone, D.G. Remick, M. LeGendre, M. McClain, and N.C. Engleberg. 1997. The role of *Legionella pneumophila*-infected *Hartmannella vermiformis* as an infectious particle in a murine model of legionnaires' disease. *Infect. Immun.* 65:5330-5333.

Brown, E., R. Tyndall, R. MacKowski, D. Hollis, R. Weyant, J. Pruckler, R. Benson, T. Messner, and B. Fields. 1997. DNA from *Bacillus laterosporus* amplifies with the Perkin Elmer EnviroAmp *Legionella* Kit. Presented at 97th General Meeting American Society for Microbiology, Miami Beach, Florida, May 4-8. American Society for Microbiology.

Bueschung, W. J., R. A. Brust, and L.W. Ayers. 1983. Enhanced primary isolation of *Legionella pneumophila* from clinical specimens by low-pH treatment. *J. Clin. Microbiol.* 17:1153-1155.

Butler, J. C., R. F. Benson, W. J. Drozanski, T. J. Rowbotham, I. Bialkowska, and D. Losos. 1995. Pneumonia associated with *Legionella*-like amoebal pathogen (LLAP) infection: Patient characteristics and clinical manifestations. *35th Interscience Conference on Antimicrobial Agents and Chemotherapy,* San Francisco, California, USA: American Society for Microbiology.

Byrd, J. J., H.-S. Xu, and R.R. Colwell. 1991. Viable but nonculturable bacteria in drinking water. *Appl. Environ. Microbiol.* 57:875-878.

Byrd, T. F., and M. A. Horwitz. 1987. Intracellular multiplication of *Legionella pneumophila* in human monocytes is iron-dependent and the capacity of activated monocytes to inhibit intracellular multiplication is reversed by iron-transferrin. *Clin. Res.* 35:613.

Byrd, T. F., and M. A. Horwitz. 1989. Interferon gamma-activated human monocytes downregulate transferrin receptors and inhibit the intracellular multiplication of *Legionella pneumophila* by limiting the availability of iron. *J. Clin. Invest.* 83:1457-1465.

Campbell, J., W. F. Bibb, M. A. Lambert, S. Eng, A. G. Steigerwalt, J. Allard, C. W. Moss, and D. J. Brenner. 1984. *Legionella sainthelensi*: A new species of *Legionella* isolated from water near Mt. St. Helens. *Appl. Environ. Microbiol.* 47:369-373.

Castellani-Pastoris, M., L. Ciceroni, R. Lo Monaco, P. Goldoni, B. Mentore, G. Flego, L. Cattani, S. Ciarrocchi, A. Pinto, and P. Visca. 1997. Molecular epidemiology of an outbreak of legionnaires' disease associated with a cooling tower in Genova-Sestri Ponente, Italy. *Eur. J. Clin. Microbiol. Infect. Dis.* 16:883-892.

Catalan, V., F. Garcia, C. Moreno, M.J. Vila, and D. Apraiz. 1997. Detection of *Legionella pneumophila* in waste water by nested polymerase chain reaction. *Res. Microbiol.* 148:71-78.

Catalan, V., C. Moreno, M.A. Dasi, C. Munoz, and D. Apraiz. 1994. Nested polymerase chain reaction for detection of *Legionella pneumophila* in water. *Res. Microbiol.* 145:603-610.

Charlotte Smith and Associates, Inc. 1997. *Assessment of Coliform Occurrence in the Water Utility A Wellfield.*

Cherry, W. B., G. W. Gorman, L.H. Orrison, C.W. Moss, A.G. Steigerwalt, H.W. Wilkinson, S.E. Johnson, R.M. McKinney, and D.J. Brenner. 1982. *Legionella jordanis*: A new species of *Legionella* isolated from water and sewage. *J. Clin. Microbiol.* 15:290-297.

Cianciotto, N. P. 2001. Pathogenicity of *Legionella pneumophila*. *Int. J. Med. Microbiol.* 291(5):331-43.

Cirillo, J. D., S. Falkow, and L.S. Tompkins. 1994. Growth of *Legionella pneumophila* in *Acanthamoeba castellanii* enhances invasion. *Infect. Immun.* 62:3254-3261.

Cullimore, D.R. 1992. *Practical Manual of Groundwater Microbiology.*

Dennis, P. J., D. J. Brenner, W. L. Thacker, R. Wait, G. Vesey, A. G. Steigerwalt, and R. F. Benson. 1993. Five new *Legionella* species isolated from water. *Int. J. Syst. Bacteriol.* 43:329-337.

Dennis, P. J., D. Green, and B.P. Jones. 1984. A note on the temperature tolerance of *Legionella. J. Appl. Bacteriol.* 56:349-350.

Dennis, P. J., J. A. Taylor, and G.I. Barrow. 1981. Phosphate buffered, low sodium chloride blood agar medium for *Legionella pneumophila. Lancet* 2(8247):636.

Dowling, J. N., F. J. Kroboth, M. Karpf, R.B. Yee, and A.W. Pasculle. 1983. Pneumonia and multiple lung abscesses caused by dual infection with *Legionella micdadei* and *Legionella pneumophila. Am. Rev. Respir. Dis.* 127:121-125.

Dowling, J. N., A. W. Pasculle, F.N. Frola, M.K. Zaphyr, and B.B Yee. 1984. Infections caused by *Legionella pneumophila* and *Legionella micdadei* among renal transplant recipients. *J. Infect. Dis.* 149:703-713.

Drozanski, W. J. 1991. *Sarcobium lyticum* gen. nov., sp. nov., an obligate intracellular bacterial parasite of small free-living amoebae. *Int. J. Syst. Bacteriol.* 41:82-87.

Dumais-Pope, C., W. O'Connell, and N.P. Cianciotto. 1996. *Legionella pneumophila* mutants that are defective for iron acquisition and assimilation and intracellular infection. *Infect. Immun.* 64:629-636.

Edelstein, P. H. 1982. Comparative study of selective media for isolation of *Legionella pneumophila* from potable water. *J. Clin. Microbiol.* 16(4):697-699.

Edelstein, P. H. 1993. Legionnaires' disease. *Clin. Infect. Dis.* 16:741-749.

Edelstein, P. H., W. F. Bibb, G. W. Gorman, W. L. Thacker, D. J. Brenner, H. W. Wilkinson, C. W. Moss, R. S. Buddington, C. J. Dunn, P. J. Roos, and P. L. Meenhorst. 1984. *Legionella pneumophila* serogroup 9: A cause of human pneumonia. *Ann. Intern. Med.* 101:196-198.

Edelstein, P. H., D. J. Brenner, C. W. Moss, A. G. Steigerwalt, E. M. Francis, and W. L. George. 1982. *Legionella wadsworthii* species nova: A cause of human pneumonia. *Ann. Intern. Med.* 97:809-813.

Edelstein, P. H., and R. D. Meyer. 1994. *Legionella* pneumonias. In *Respiratory Infections: Diagnosis and Management,* 3rd ed. Edited by J. E. Pennigton. New York: Raven Press, Ltd. 455-484.

Edelstein, P. H., J. B. Snitzer, and J.A. Bridge. 1982. Enhancement of recovery of *Legionella pneumophila* from contaminated respiratory tract specimens by heat. *J. Clin. Microbiol.* 16:1061-1065.

Edelstein, P. H., J. B. Snitzer, and S. M. Finegold. 1982. Isolation of *Legionella pneumophila* from hospital potable water specimens: Comparison of direct plating with guinea pig inoculation. *J. Clin. Microbiol.* 15:1092-1096.

England, A. C., R. M. McKinney, P. Skaliy, and G. W. Gorman. 1980. A fifth serogroup of *Legionella pneumophila. Annals of Internal Medicine* 93(1):58-59.

Ezzedine, K., C. Van Ossel, M. Delmee, and G. Wauters. 1989. *Legionella* spp. in a hospital hot water system: Effect of control measures. *J. Hosp. Infect.* 13:121-131.

Fang, G. D., J. E. Stout, V.L. Yu, A. Goetz, J.D. Rihs, and R.M. Vickers. 1990. Community-acquired pneumonia caused by *Legionella dumoffii* in a patient with hairy cell leukemia. *Infection* 18:383-385.

Fang, G.-D., V. L. Yu, and R. M. Vickers. 1989. Disease due to the Legionellaceae (other than *Legionella pneumophila*). Historical, microbiological, clinical, and epidemiological review. *Medicine* 68(2):116-132.

Feeley, J. C., R. J. Gibson, G.W. Gorman, N.C. Langford, J.K. Rasheed, D.C. Mackel, and W.B. Baine. 1979. Charcoal yeast extract agar: Primary isolation medium for *Legionella pneumophila*. *J. Clin. Microbiol.* 10:437-441.

Feeley, J. C., G. W. Gorman, R.E. Weaver, D.C. Mackel, and H.W. Smith. 1978. Primary isolation media for legionnaires' disease bacterium. *J. Clin. Microbiol.* 8(3):320-325.

Felsenstein, J. 1981. Evolutionary trees from DNA sequences: A maximum likelihood approach. *J. Molecular Evolution* 17:368-376.

Fields, B. S. 1993. *Legionella* and Protozoa: Interaction of a pathogen and its natural host. In *Legionella: Current Status and Emerging Perspectives.* Edited by J. M. Barbaree, R. F. Breiman, and A. P. Dufour. Washington, D.C.: American Society for Microbiology. 129-136.

Fields, B. S. 1996. The molecular ecology of Legionellae. *Trends in Microbiology* 4:286-290.

Fields, B. S. 1997. Legionellae and legionnaires' disease. In *Manual of Environmental Microbiology.* Edited by C. J. Hurst, G. R. Knudsen, M. J. McInerney, L. D. Stetzenbach, and M. V. Walter. Wasington, D.C.: American Society for Microbiology. 666-675.

Fields, B. S., J. M. Barbaree, G.N. Sanden, and W.E. Morrill. 1990. Virulence of a *Legionella anisa* strain associated with Pontiac fever: An evaluation using protozoan, cell culture, and guinea pig models. *Infect. Immun.* 58:3139-3142.

Fields, B. S., S. R. Fields, J. N. Loy, E. H. White, W. L Steffens, and E. B. Shotts. 1993. Attachment and entry of *Legionella pneumophila* in *Hartmannella vermiformis*. *J. Infect. Dis.* 167:1146-1150.

Fields, B. S., G. N. Sanden, J. M. Barbaree, W. E. Morrill, R. M. Wadowsky, and E. H. White. 1989. Intracellular multiplication of *Legionella pneumophila* in amoebae isolated from hospital hot water tanks. *Curr. Microbiol.* 18:131-137.

Fields, B. S., E. B. Shotts, J. C. Feeley, G. W. Gorman, and W. T. Martin. 1984. Proliferation of *Legionella pneumophila* as an intracellular parasite of the ciliated protozoan *Tetrahymena pyriformis*. *Appl. Environ. Microbiol.* 47:467-471.

Fitch W. M., and E. Margoliash. 1967. Construction of phylogenetic trees. *Science* 155(760): 279-284.

Fitzgeorge, R. B., and P. J. Dennis. 1983. Isolation of *Legionella pneumophila* from water supplies: Comparison of methods based on the guinea-pig and culture media. *J. Hyg.* 91:179-186.

Fliermans, C. B. 1983. Autecology of *Legionella pneumophila*. *Zbl. Bakt. Hyg.* 255:58-63.

Fliermans, C. B. 1984. Philosophical ecology: *Legionella* in historical perspective. *Legionella: Proceedings of the 2nd International Symposium.* Edited by C. Thornsberry, A. Balows, J. C. Feeley, and W. Jakubowski. Washington, D. C.: American Society for Microbiology. 285-289.

Fliermans, C. B. 1996. Ecology of *Legionella*: From data to knowledge with a little wisdom. *Microbiol. Ecol.* 32: 203-228.

Fliermans, C. B., W. B. Cherry, L.H. Orrison, S.J. Smith, D.L. Tison, and D.H. Pope. 1981. Ecological distribution of *Legionella pneumophila*. *Appl. Environ. Microbiol.* 41:9-16.

Foissner, W. 1996. Ciliates in rapid gravity filters of waterworks exploiting deep groundwaters. *Microsc. Res. Tech.* 33:12-22.

Forey, F., S. Jarraud, M. Reyrolle, C. Lawrence, B. Decludt, M. Molmeret, S. Riffard, F. LoPresti, F. Vandenesch, and J. Etienne. 1998. Recognition by pulsed-field gel electrophoresis analysis of the widespread distribution of virulent strains of *Legionella pneumophil*a serogroup 1 in the general water system of different cities. Presented at *13th meeting of the European Working Group on Legionella Infections*, Helsinki, Finland.

Fox, G. E., J. D. Wisotzkey, and P. Jurtshuk. 1992. How close is close: 16S rRNA sequence identity may not be sufficient to guarantee species identity. *Int. J. Syst. Bacteriol.* 42:166-170.

Fry, N. K., T. J. Rowbotham, N.A. Saunders, and T.M. Embley. 1991a. Direct amplification and sequencing of the 16S ribosomal DNA of an intracellular *Legionella* species recovered by amoebal enrichment from the sputum of a patient with pneumonia. *FEMS Microbiol. Lett.* 83:165-168.

Fry, N. K., S. Warwick, N.A. Saunders, and T.M. Embley. 1991b. The use of 16 S ribosomal RNA analysis to investigate the phylogeny of the family Legionellaceae. *J. Gen. Microbiol.* 137:1215-1222.

Gebran, S. J., C. A. Newton, Y. Yamamoto, R. Widen, T.W. Klein, and H. Friedman. 1994. Macrophage permissiveness for *Legionella pneumophila* growth modulated by iron. *Infect. Immun.* 62:564-568.

Gilbart, J., and M. D. Collins 1985. High-performance liquid chromatographic analysis of ubiquinones from new *Legionella* species. *FEMS Microbiol. Lett.* 26:77-82.

Goldoni, P., P. Visca, M.C. Pastoris, P. Valenti, and N. Orsi. 1991. Growth of *Legionella* spp. under conditions of iron restriction. *J. Med. Microbiol.* 34:113-118.

Gomez-Lus, R., E. Lomba, P. Gomez-Lus, M. S. Abarca, S. Gomez-Lus, A. Martinez, E. Duran, and M. C. Rubio. 1993. In vitro antagonistic activity of *Pseudomonas aeruginosa*, *Klebsiella pneumoniae*, and *Aeromonas* spp. against *Legionella* spp. In *Legionella: Current Status and Emerging Perspectives*. Edited by J. M. Barbaree, R. F. Breiman, and A. P. Dufour. Washington, D.C.: American Society for Microbiology. 265-267.

Graman, P. S., G. A. Quinlan, and J.A. Rank. 1997. Nosocomial legionellosis traced to a contaminated ice machine. *Infect. Control Hosp. Epidemiol.* 18:637-640.

Griffith, M. E., D. S. Lindquist, R.F. Benson, W.L. Thacker, D.J. Brenner, and H.W. Wilkinson. 1988. First isolation of *Legionella gormanii* from human disease. *J. Clin. Microbiol.* 26:380-381.

Grimm, D., H. Merkert, W. Ludwig, K.H. Schleifer, J. Hacker, and B.C. Brand. 1998. Specific detection of *Legionella pneumophila*: Construction of a new 16S rRNA-targeted oligonucleotide probe. *Appl. Environ. Microbiol.* 64:2686-2690.

Grimont, F., M. Lefèvre, E. Ageron, and P.A. Grimont. 1989. rRNA gene restriction patterns of *Legionella* species: A molecular identification system. *Res. Microbiol.* 140:615-626.

Gürtler, V., and H. D. Barrie. 1995. Typing of *Staphylococcus aureus* strains by PCR-amplification of variable-length 16S-23S rDNA spacer regions: Characterization of spacer sequences. *Microbiology* 141:1255-1265.

Harrell, L. J., G. L. Andersen, and K.H. Wilson. 1995. Genetic variability of *Bacillus anthracis* and related species. *J. Clin. Microbiol.* 33:1847-1850.

Harrington, R. D., A. E. Woolfrey, R. Bowden, M.G. McDowell, and R.C. Hackman. 1996. Legionellosis in a bone marrow transplant center. *Bone Marrow Transplantation* 18:361-368.

Harrison, T. G., N. A. Saunders, A. Haththotuwa, G. Hallas, R.J. Birtles, and A.G. Taylor. 1990. Phenotypic variation amongst genotypically homogeneous *Legionella pneumophila* serogroup 1 isolates: Implications for the investigation of outbreaks of legionnaires' disease. *Epidemiol. Infect.* 104:171-180.

Harrison, T. G., and A. G. Taylor. 1988a. Identification of Legionellae by serological methods. In *A Laboratory Manual for* Legionella. Edited by T. G. Harrison and A. G. Taylor. Chichester: John Wiley & Sons, Ltd. 57-68.

Harrison, T. G., and A. G. Taylor. 1988b. Phenotypic characteristics of Legionellae. In *A Laboratory Manual for* Legionella. Edited by T. G. Harrison and A. G. Taylor. Chichester: John Wiley & Sons, Ltd. 45-56.

Hebert, G. A., A. G. Steigerwalt, and D. J. Brenner. 1980. *Legionella micdadei* species nova: Classification of a third species of *Legionella* associated with human pneumonia. *Curr. Microbiol.* 3:255-257.

Helbig, J. H., B. Ludwig, P.C. Luck, A. Groh, W. Witzleb, and J. Hacker. 1995. Monoclonal antibodies to *Legionella* Mip proteins recognize genus- and species-specific epitopes. *Clin. Diagn. Lab. Immunol.* 2:160-165.

Heller, R., C. Holler, R. Sussmuth, and K. Gunderman. 1998. Effect of salt concentration and temperature on survival of *Legionella pneumophila*. *Lett. Appl. Microbiol.* 26:64-68.

Hernandez, J. F., J. M. Delattre, and C. Ogre. 1983. Thermorésistance des *Legionella*. *Ann. Microbiol. (Inst. Pasteur)* 134(B):421-427.

Herwaldt, L. A., G. W. Gorman, T. McGrath, S. Toma, B. Brake, A. W. Hightower, J. Jones, A. L. Reingold, P. A. Boxer, P. W. Tang, C. W. Moss, H. W. Wilkinson, D. J. Brenner, A. G. Steigerwalt, and C. V. Broome. 1984. A new *Legionella* species, *Legionella feeleii* species nova, causes Pontiac fever in an automobile plant. *Ann. Int. Med.* 100:333-338.

Hookey, J. V., N. A. Saunders, N. K. Fry, R. J. Birtles, and T. G. Harrison. 1996. Phylogeny of Legionellaceae based on small-subunit ribosomal DNA sequences and proposal of *Legionella lytica* comb. nov. for *Legionella*-like amoebal pathogens. *Int. J. Syst. Bacteriol.* 46:526-531.

Horbach, I., D. Naumann, and F.J. Fehrenbach. 1988. Simultaneous infections with different serogroups of *Legionella pneumophila* investigated by routine methods and Fourier transform infrared spectroscopy. *J. Clin. Microbiol.* 26:1106-1110.

Hume, R. D., and W. D. Hann. 1984. Growth relationships of *Legionella pneumophila* with green algae (*Chlorophyta*). *Legionella: Proceedings of the 2nd International Symposium, Atlanta, Georgia.* Washington, D.C.: American Society for Microbiology. 323-324.

Hussong, D., R. R. Colwell, M. O'Brien, E. Weiss, A. D. Pearson, R. M. Weiner, and W. D. Burge. 1987. Viable *Legionella pneumophila* not detectable by culture on agar media. *Bio/Technology* 5:947-950.

Infuso, A., B. Hubert, and J. Etienne. 1998. La sous-déclaration de la légionellose en France: Pour une surveillance plus active. *Eurosurveillance* 3:48-50.

Jalaludin, B. B., O. T. K. Nguyen, I.C. Goldthorpe, and R.F. Chiew. 1997. Legionnaires' disease: co-infection with *Legionella pneumophila* serogroups 1 and 5. *Med. J. Aust.* 166:277-278.

James, B. W., W. S. Mauchline, P.J. Dennis, C.W. Keevil, and R. Wait. 1999. Poly-3-hydroxybutyrate in *Legionella pneumophila*, an energy source for survival in low-nutrient environment. *Appl. Environ. Microbiol.* 65:822-827.

James, B. W., W. S. Mauchline, R.B. Fitzgeorge, P.J. Dennis, and C.W. Keevil. 1995. Influence of iron-limited continuous culture on physiology and virulence of *Legionella pneumophila*. *Infect. Immun.* 63:4224-4230.

Jaulhac, B., C. Harf, M. Nowicki, and H. Monteil. 1993. Detection of *Legionella* spp. in environmental water samples and free-living amoeba by using DNA amplification. In *Legionella: Current Status and Emerging Perspectives*. Edited by J. M. Barbaree, R. F. Breiman, and A. P. Dufour. Washington, D.C.: American Society for Microbiology. 151-153.

Jaulhac, B., M. Nowicki, N. Bornstein, O. Meunier, G. Prevost, Y. Piemont, J. Fleurette, and H. Monteil. 1992. Detection of *Legionella* spp. in bronchoalveolar lavage fluids by DNA amplification. *J. Clin. Microbiol.* 30:920-924.

Jensen, M. A., J. A. Webster, and N. Straus. 1993. Rapid identification of bacteria on the basis of polymerase chain reaction-amplified ribosomal DNA spacer polymorphisms. *Appl. Environ. Microbiol.* 59:945-952.

Johnson, S. R., W. O. Schalla, K.H. Wong, and G.H. Perkins. 1982. Simple, transparent medium for study of Legionellae. *J. Clin. Microbiol.* 15(2):342-344.

Joseph, C., and E. J. Hutchinson. 1996. Legionnaires' disease in Europe–1995. Presented at 11th Meeting of the European Working Group on Legionella Infections, Oslo, Norway. The Norwegian Defence Microbiological Laboratory, Joint Medical Services, HQ Defence Command Norway.

Kalmbach, S., W. Manz, and U. Szewzyk. 1997. Isolation of new bacterial species from drinking water biofilms and proof of their in situ dominance with highly specific 16S rRNA probes. *Appl. Environ. Microbiol.* 63:4164-4170.

Kessler, H. H., F. F. Reinthaler, A. Pschaid, K. Pierer, B. Kleinhappl, E. Eber, and E. Marth. 1993. Rapid detection of *Legionella* species in bronchoalveolar lavage fluids with the EnviroAmp *Legionella* PCR amplification and detection kit. *J. Clin. Microbiol.* 31(12):3325-3328.

Kioski, C., G. Cage, B. Johnson, C. Rosales, B. England, and T. J. Halpin. 1997. Sustained transmission of nosocomial legionnaires' disease. *Morbid. Mortal. Weekly Rep.* 46:416-421.

Koide, M., A. Saito, M. Okazaki, B. Umeda, and R. F. Benson. 1999. Isolation of *Legionella longbeachae* serogroup 1 from potting soils in Japan. *Clin. Infect. Dis.* 29:943-944.

Kool, J. L., J. C. Carpenter, and B. S. Fields. 1999. Effect of monochloramine disinfection of municipal drinking water on risk of nosocomial Legionnaires' disease. *Lancet* 353:272-277.

Kramer, M. H. J. and T. E. Ford 1994. Legionellosis: Ecological factors of an environmentally new disease. *Zbl. Hyg.* 195:470-482.

Kuchta, J. M., J. S. Navratil, M. E. Shepherd, R. M. Wadowsky, J. N. Dowling, S. J. States, and R. B. Yee 1993. Impact of chlorine and heat on the survival of *Hartmannella vermiformis* and subsequent growth of *Legionella pneumophila*. *Appl. Environ. Microbiol.* 59(12):4096-4100.

Kuchta, J. M., S. J. States, J. E. McGlaughlin, J. H. Overmeyer, R. M. Wadowsky, A. M. McNamara, R. S. Wolford, and R. B. Yee. 1984. Enhanced chlorine resistance of tap water adapted *Legionella pneumophila* as compared with agar medium passed strains. *Appl. Environ. Microbiol.* 50:21-26.

Lambert, M. A., and W. C. Moss. 1989. Cellular fatty acid compositions and isoprenoid quinone contents of 23 *Legionella* species. *J. Clin. Microbiol.* 27(3):465-473.

LeChevallier, M. W., W. Schulz, and R.G. Lee. 1991. Bacterial nutrients in drinking water. *Appl. Environ. Microbiol.* 57:857-862.

Lee, D.-H., Y.-G. Zo, and S.J. Kim. 1996. Nonradioactive method to study genetic profiles of natural bacterial communities by PCR-single-strand-conformation polymorphism. *Appl. Environ. Microbiol.* 62:3112-3120.

Lee, T. C., R. M. Vickers, V.L. Yu, and M.M. Wagener. 1993. Growth of 28 *Legionella* species on selective culture media: A comparative study. *J. Clin. Microbiol.* 31:2764-2768.

Leoni, E., and P. P. Legnani. 2001. Comparison of selective procedures for isolation and enumeration of *Legionella* species from hot water systems. *J. Appl. Microbiol.* 90:27-33.

Lin, Y., R. Vidic, J. E. Stout, and V. L. Yu. 1992. *Legionella* in water distribution systems. *Waterborne Pathogens* 90:112-122.

Lin, Y. E., R. D. Vidic, J. E. Stout, and V. L. Yu. 1998. *Legionella* in water distribution systems. *Jour. AWWA* 90(9):112-121.

Lin, Y. S. E., R. D. Vidic, J. E. Stout, and V. L. Yu. 1996. Individual and combined effects of copper and silver ions on inactivation of *Legionella pneumophila*. *Water Res.* 30(8):1905-1913.

Lin, Y. S. E., R. D. Vidic, J. E. Stout, and V. L. Yu. 2002. Negative effect of high pH on biocidal efficacy of copper and silver ions in controlling *Legionella pneumophila*. *Appl. Environ. Microbiol.* 68(6):2711-2715.

Lindquist, D. S., G. Nygaard, W. L. Thacker, R. F. Benson, D. J. Brenner, and H. W. Wilkinson. 1988. Thirteenth serogroup of *Legionella pneumophila* isolated from patients with pneumonia. *J. Clin. Microbiol.* 26:586-587.

Liu, Z., J. E. Stout, M. Boldin, J. Rugh, W. F. Diven, and V. L. Yu. 1998. Intermittent use of copper-silver ionization for *Legionella* control in water distribution systems: A potential option in buildings housing individuals at low risk infection. *Clin. Infect. Dis.* 26:138-140.

Lo Presti, F., M. Molmeret, S. Riffard, S. Jarraud, M. Reyrolle, F. Vandenesch, and J. Etiene. 1998a. Atypical cultural characteristics of *Legionella* isolates: Implication for bacterial diagnosis. Presented at 13th Meeting of the European Working Group on *Legionella* Infections, Helsinki, Finland.

Lo Presti, F., M. Reyrolle, N. Bornstein, J. Fleurette, F. Vandenesch, and J. Etienne. 1996. Three new species, one new subspecies and one new serogroup in the Legionellae. Presented at 11th Meeting of the European Working Group on *Legionella* Infections, Olso, Norway. The Norwegian Defence Microbiological Laboratory, Joint Medical Services HQ Defence Command Norway.

Lo Presti, F., S. Riffard, S. Jarraud, F. Le Gallou, H. Richet, F. Vandenesch, and J. Etienne. 2000. Isolation of *Legionella oakridgensis* from two patients with pleural effusion living in the same geographical area. *J. Clin. Microbiol.* 38:3128-3130.

Lo Presti, F., S. Riffard, H. Meugnier, M. Reyrolle, Y. Lasne, P. A. Grimont, F. Grimont, R. F. Benson, D. J. Brenner, A. G. Steigerwalt, J. Etienne, and J. Freney. 2001. *Legionella gresilensis* sp. nov., and *Legionella beliardensis* sp. nov., two new species isolated from water in France. *Int. J. Syst. Evol. Microbiol.* 51:1949 -1957.

Lo Presti, F., S. Riffard, H. Meugnier, M. Reyrolle, Y. Lasne, P.A. Grimont, F. Grimont, F. Vandenesch, J. Etienne, J. Fleurette, and J. Freney. 1999. *Legionella taurinensis*, a new species antigenically similar to *Legionella spiritensis*. *Int. J. Syst. Bacteriol.* 49:397-403.

Lo Presti, F., S. Riffard, F. Vandenesch, and J. Etienne. 1998b. Identification of *Legionella* species by random amplified polymorphic DNA profiles. *J. Clin. Microbiol.* 38:3193-3197.

Lo Presti, F., S. Riffard, F. Vandenesch, M. Reyrolle, E. Ronco, P. Ichai, and J. Etienne. 1997. The first clinical isolate of *Legionella parisiensis*, from a liver transplant patient with pneumonia. *J. Clin. Microbiol.* 35:1706-1709.

Loutit, J. S., and L. S. Tompkins. 1993. PCR detection of *Legionella pneumophila* and *L.dumoffii* in water. In *Diagnostic Molecular Microbiology: Principles and Applications*. Edited by D. H. Persing, T. F. Smith, F. C. Tenover, and T. J. White. Washington, D.C.: American Society for Microbiology. 261-265.

Lück, P. C., H.-M. Wenchel, and J.H. Helbig. 1998. Nosocomial pneumonia caused by three genetically different strains of *Legionella pneumophila* and detection of these strains in the hospital water supply. *J. Clin. Microbiol.* 36:1160-1163.

Lye, D., G. S. Fout, S. R. Crout, R. Danielson, C. L. Thio, and C. M. Paszko-Kolva. 1997. Survey of ground, surface, and potable waters for the presence of *Legionella* species by EnviroAmp® PCR *Legionella* kit, culture, and immunofluorescent staining. *Water Res.* 31:287-293.

Ma, P., G. S. Visvesvara, A.J. Martinez, F.H. Theodore, P.M. Daggett, and T.K. Sawyer. 1990. *Naegleria* and *Acanthamoeba* infections. *Rev. Infect. Dis.* 12:490-513.

Maiwald, M., M. Schill, C. Stockinger, J.H. Helbig, P.C. Luck, W. Witzleb, and H.G. Sonntag. 1995. Detection of *Legionella*-DNA in human and guinea pig urine samples by the polymerase chain reaction. *Eur. J. Clin. Microbiol.* 14:25-33.

Makin, T., and C. A. Hart. 1993. Efficacy of UV radiation for eradicating *Legionella pneumophila* from a shower. In *Legionella: Current Status and Emerging Perspectives.* Edited by J. M. Barbaree, R. F. Breiman, and A. P. Dufour. Washington, D.C.: American Society for Microbiology. 255-258.

Marmet, D., N. Bornstein, and J. Fleurette. 1988. Identification des *Legionella* par analyse des acides gras en chromatographie phase gazeuse (CPG) et des ubiquinones en chromatographie liquide haute performance (CLHP). *Ann. Biol. Clin.* 46:371-375.

Marrie T, P. Green, S. Burbridge, G. Bezanson, S. Neale, P. S. Hoffman, and D. Haldane. 1994. Legionellaceae in the potable water of Nova Scotia hospitals and Halifax residences. *Epidemiol. Infect.* 112(1):143-50.

Marrie, T. J., D. Raoult, B. La Scola, R. J. Birtles, and E. de Carolis; Canadian Community-Acquired Pneumonia Study Group. 2001. *Legionella*-like and other amoebal pathogens as agents of community acquired pneumonia. *Emerging Infectious Diseases* 7(6):1026-1029.

Marshall, M. M., D. Naumovitz, Y. Ortega, and C.R. Sterling. 1997. Waterborne protozoan pathogens. *Clin. Microbiol. Rev.* 10:67-85.

Martin, W. T., B. S. Fields, and L. C. Hutwagoner. 1993. Comparison of culture and polymerase chain reaction to detect Legionellae in environmental samples. In *Legionella: Current Status and Emerging Perspectives*. Edited by J. M. Barbaree, R. F. Breiman, and A. P. Dufour. Washington, D.C.: American Society for Microbiology. 175-176.

Matsiota-Bernard, P., E. Pitsouni, N. Legakis, and C. Nauciel. 1994. Evaluation of commercial amplification kit for detection of *Legionella pneumophila* in clinical specimens. *J. Clin. Microbiol.* 32:1503-1505.

Mauchline, W. S., and C. W. Keevil. 1991. Development of the BIOLOG substrate utilization system for identification of *Legionella* spp. *Appl. Environ. Microbiol.* 57:3345-3349.

McKinney, R., L. Thacker, P. P Harris, K. R. Lewallen, G. A. Hebert, P. H. Edelstein, and B. M. Thomason. 1979. Four serogroups of Legionnaires' disease bacteria defined by direct immunofluorescence. *Ann. Intern. Med.* 90:621-624.

McKinney, R. M., R. K. Porschen, P. H. Edelstein, M. L. Bissett, P. P. Harris, S. P. Bondell, A. G. Steigerwalt, R. E. Weaver, M. E. Ein, D. S. Lindquist, R. S. Kops, and D. J. Brenner. 1981. *Legionella longbeachae* species nova, another etiologic agent of human pneumonia. *Ann. Int. Med.* 94:739-743.

McKinney, R. M., H. W. Wilkinson, H. M. Sommers, B. J. Fikes, K. R. Sasseville, M. M. Yungbluth, and J. S. Wolf. 1980. *Legionella pneumophila* serogroup six: Isolation from cases of legionellosis, identification by immunofluorescence staining, and immunological response to infection. *J. Clin. Microbiol.* 12:395-401.

Meenhorst, P. L., A. L. Reingold, G. W. Gorman, J. C. Feeley, B. J. van Cronenburg, C. L. Meyer, and R. van Furth. 1983. *Legionella* pneumonia in guinea pigs exposed to aerosols of concentrated potable water from a hospital with nosocomial legionnaires' disease. *J. Infect. Dis.* 147:129-132.

Meenhorst, P. L., A. L. Reingold, D. G. Groothuis, G. W. Gorman, H. W. Wilkinson, R. M. McKinney, J. C. Feeley, D. J. Brenner, and R. van Furth. 1985. Water-related nosocomial pneumonia caused by *Legionella pneumophila* serogroups 1 and 10. *J. Infect. Dis.* 152:356-364.

Meunier, J.-R., and P. A. D. Grimont. 1993. Factors affecting reproducibility of random amplified polymorphic DNA fingerprinting. *Res. Microbiol.* 144:373-379.

Miller, L. A., J. L. Beebe, J.C. Butler, W. Martin, R. Benson, R.E. Hoffman, and B.S. Fields. 1993. Use of polymerase chain reaction in an epidemiologic investigation of Pontiac fever. *J. Infect. Dis.* 168:769-772.

Miragliotta, G., R. Del Prete, R. Sabato, A. Cassano, and N. Carnimeo. 1992. Legionellosis associated with artesian well excavation. *Eur. J. Epidemiol.* 8:748-749.

Mitchell, K., and R. Fallon. 1990. The determination of ubiquinones profiles by reversed-phase high-performance thin-layer chromatography as an aid to the speciation of Legionellaceae. *J. Gen. Microbiol.* 136:2035-2041.

Miyamoto, H., H. Yamamoto, K. Arima, J. Fujii, K. Maruta, K. Izu, T. Shiomori, and S. Yoshida. 1997. Development of a new seminested PCR method for detection of *Legionella* species and its application to surveillance of Legionellae in hospital cooling tower water. *Appl. Environ. Microbiol.* 63:2489-2494.

Molmeret, M., S. Jarraud, P. Pernin, P. Farge, G. Lina, F. Vandenesch, and J. Etienne. 2000. *Legionella pneumophila* replication following uptake by *Hartmannella* and subsequent phagocytosis by *Willaertia.* Presented at 5th International Conference on *Legionella*, September 26-29, 2000. Ulm, Germany.

Morgan, J. A. W., C. Winstanley, R. W. Pickup, and J. R. Saunders. 1991. Rapid immunocapture of *Pseudomonas putida* cells from lake water by using bacterial flagella. *Appl. Environ. Microbiol.* 57:503-509.

Morris, G. K., A. G. Steigerwalt, J.C. Feeley, E.S. Wong, W.T. Martin, C.M. Patton, and D.J. Brenner. 1980. *Legionella gormanii* sp. nov. *J. Clin. Microbiol.* 12:718-721.

Moss, C. W., W. F. Bibb, D.E. Karr, G.O. Guerrant, and M.A. Lambert. 1983. Cellular fatty acid composition and ubiquinone content of *Legionella feeleii* sp. nov. *J. Clin. Microbiol.* 18(4):917-919.

Murray, R. G. E., D. J. Brenner, R. R. Colwell, P. deVos, M. Goodfellow, P. A. D. Grimont, N. Pfenning, E. Stackelbrandt, and G. D. Zavarzin. 1990. Report of the ad hoc committee on approaches to taxonomy within the Proteobacteria. *Int. J. Syst. Bacteriol.* 40:213-215.

Muyzer, G. 1999. DGGE/TGGE a method for identifying genes from natural ecosystems. *Curr. Opinion Microbiol.* 2:317-322.

Muyzer, G., T. Brinkhoff, U, Nübel, C. Santegoeds, H. Schäfer, and C. Wawer. 1998. Denaturing gradient gel electrophoresis (DGGE) in microbial ecology. *Molecular Microbiology Ecology Manual,* 3rd ed. Edited by A. D. L. Akkermans, J. D. van Elsas, and F. J. de Bruijn. Dordrecht, The Netherlands: Kluwer Academic Publishers. 3.4.4:1-27.

Muyzer, G., E. C. De Waal, and A.G. Uitterlinden. 1993. Profiling of complex microbial populations by denaturing gradient gel electrophoresis analysis of polymerase chain reaction-amplified genes coding for 16S rRNA. *Appl. Environ. Microbiol.* 59:695-700.

Muyzer, G., S. Hottenträger, A. Teske, and C. Wawer. 1995. Denaturing gradient gel electrophoresis of PCR-amplified 16S rDNA-a new molecular approach to analyse the genetic diversity of mixed microbial communities. In *Molecular Microbial Ecology Manual,* 2nd ed. Edited by A. D. L. Akkermans, J. D. van Elsas, and F. J. de Bruijn. Dordrecht, The Netherlands: Kluwer Academic Publishers.

Muyzer, G., and K. Smalla. 1998. Application of denaturing gradient gel electrophoresis (DGGE) and temperature gradient gel electrophoresis (TGGE) in microbial ecology. *Antonie van Leeuwenhoek* 73:127-141.

Nahapetian, K., O. Challemel, D. Beurtin, S. Dubrou, P. Gounon, and F. Squinazi. 1991. The intracellular multiplication of *Legionella pneumophila* in protozoa from hospital plumbing system. *Res. Microbiol.* 142:677-685.

Newsome, A. L., T. M. Scott, R.F. Benson, and B.S. Fields. 1998. Isolation of an amoeba naturally harboring a distinctive *Legionella* species. *Appl. Environ. Microbiol.* 64:1688-1693.

Ng, D. K. L., B. B. Koh, L. Tay, and B.H. Heng. 1997. Comparison of polymerase chain reaction and conventionnal culture for the detection of legionellae in cooling tower waters in Singapore. *Lett. Appl. Microbiol.* 24:214-216.

Normand, P., B. Cournoyer, P. Simonet, and S. Nazaret. 1992. Analysis of a ribosomal RNA operon in the actinomycete *Frankia*. *Gene* 111:119-124.

Normand, P., C. Ponsonnet, X. Nesme, M. Neyra, and P. Simonet. 1996. ITS analysis of prokaryotes. In *Molecular Microbial Ecology Manual.* Edited by A. D. L. Akkermans, J. D. van Elsas, and F. J. De Bruijn. Dordrecht, The Netherlands: Kluwer Academic Publishers. 3.4.5:1-12.

Nowicki, M., N. Bornstein, B. Jaulhac, Y. Piemont, H. Monteil and J. Fleurette. 1993. Rapid detection of Legionellae in clinical and environmental samples by polymerase chain reaction. In *Legionella: Current Status and Emerging Perspectives*. Edited by J. M. Barbaree, R. F. Breiman, and A. P. Dufour. Washington, D.C.: American Society for Microbiology. 178-181.

Nowicki, M., N. Bornstein, J.C. Paucod, P. Binder, and J. Fleurette. 1987. Effect of culture medium on morphology and virulence of *Legionella pneumophila* serogroup 1. *Zentralblatt für Bakteriologie Mikrobiologie und Hygiene* 264:167-177.

O'Connell, W. A., J. M. Bangsborg, and N.P. Cianciotto. 1995. Characterization of a *Legionella micdadei* mip mutant. *Infect. Immun.* 63:2840-2845.

O'Connell, W. A., E. K. Hickey, and N.P. Cianciotto. 1996. A *Legionella pneumophila* gene that promotes hemin binding. *Infect. Immun.* 64:842-848.

Okpara, J., M. Maiwald, M. Borneff, J. Windeler, and H.G. Sonntag. 1996. Evaluation of a new version of the EnviroAmp *Legionella* kit for the detection of legionellae in water samples by the polymerase chain reaction. *Zbl. Hyg.* 198:502-513.

Olsen, G. J., R. Overbeek, N. Larsen, T. L. Marsh, M. J. McCaughey, M. A. Maciukenas, W. M. Kuan, T. J. Macke, Y. Xing, and C. R. Woese. 1992. The ribosomal database project. *Nucleic Acids Res.* 20:2199-2200.

Olsvik, O., T. Popovic, E. Skjerve, K.S. Cudjoe, E. Hornes, J. Ugelstad, and M. Uhlen. 1994. Magnetic separation techniques in diagnostic microbiology. *Clin. Microbiol. Rev.* 7: 43-54.

Orrison, L. H., W. B. Cherry, R.L. Tyndall, C.B. Fliermans, S.B. Gough, M.A. Lambert, L. K. McDougal, W.F. Bibb, and D.J. Brenner. 1983. *Legionella oakridgensis*: Unusual new species isolated from cooling tower water. *Appl. Environ. Microbiol.* 45:536-545.

Palmer, C. J., G. F. Bonilla, B. Roll, C. Paszko-Kolva, L.R. Sangermano, and R.S. Fujioka. 1995. Detection of *Legionella* species in reclaimed water and air with the EnviroAmp *Legionella* PCR kit and direct fluorescent antibody staining. *Appl. Environ. Microbiol.* 61:407-412.

Palmer, C. J., Y.-L. Tsai, C. Paszko-Kolva, C. Mayer, and L.R. Sangermano. 1993. Detection of *Legionella* species in sewage and ocean water by polymerase chain reaction, direct fluorescent-antibody, and plate culture methods. *Appl. Environ. Microbiol.* 59:3618-3624.

Pankhurst, C. L., J. N. Philpott-Howard, J. H. Hewitt, and M. W. Casewell. 1990. The efficacy of chlorination and filtration in the control and eradication of *Legionella* from dental chair watersystems. *J. Hosp. Infect.* 16:9-18 (Erratum, 16:393).

Paszko-Kolva, C., P. A. Hacker, M. H. Stewart, and R. L. Wolfe. 1993a. Inhibitory effect of heterotrophic bacteria on the cultivation of *Legionella dumoffii*. In *Legionella: Current Status and Emerging Perspectives.* Edited by J. M. Barbaree, R. F. Breiman, and A. P. Dufour. Washington, D.C.: American Society for Microbiology. 203-205.

Paszko-Kolva, C., M. Shahamat, and R. R. Colwell. 1992. Long-term survival of *Legionella pneumophila* serogroup 1 under low-nutrient conditions and associated morphological changes. *FEMS Microbiol. Ecol.* 102:45-55.

Paszko-Kolva, C., Y. Yamamoto, M. Shahamat, and R. R. Colwell. 1993b. Polymerase chain reaction, gene probe, and direct fluorescent antibody staining of *Legionella pneumophila* serogroup 1 in drinking water and environmental samples. In *Legionella: Current Status and Emerging Perspectives.* Edited by J. M. Barbaree, R. F. Breiman, and A. P. Dufour. Washington, D.C.: American Society for Microbiology. 181-183.

Patterson, W. J., J. Hay, D. V. Seal, and J. C. McLuckie. 1997. Colonization of transplant unit water supplies with *Legionella* and protozoa: Precautions required to reduce the risk of legionellosis. *J. Hosp. Infect.* 37:7-17.

Patterson, W. J., D. V. Seal, E. Curran, T. M. Sinclair, and J. C. McLuckie. 1994. Fatal nosocomial Legionnaires' disease: Relevance of contamination of hospital water supply by temperature-dependent buoyancy-driven flow from spur pipes. *Epidemiol. Infect.* 112:513-525.

Picard, C., C. Ponsonnet, E. Paget, X. Nesme, and P. Simonet. 1992. Detection and enumeration of bacteria in soil by direct DNA extraction and polymerase chain reaction. *Appl. Environ. Microbiol.* 58:2717-2722.

Pietsch, M., U. Graeber, and J. Lorenz. 1995. Pneumonia caused by *Legionella feeleii*, serogroup 2: first case report from Germany. *Hyg. Med.* 20: 235-241.

Pine, L., J. R. George, M.W. Reeves, and W.K. Harrell. 1979. Development of a chemically defined liquid medium for growth of *Legionella pneumophila. J. Clin. Microbiol.* 9:615-626.

Pongratz A., A. Schwarzkopf, H. Hahn, J. Heesemann, H. Karch, and W. Doll. 1994. Impact of material of water supply systems on prevalence of *Legionella* sp. in an hospital. *Zentralblatt fur Hygiene und Umweltmedizin.* 195(5-6):483-488.

Ramirez, J. A., S. Ahkee, A. Tolentino, R.D. Miller, and J.T. Summersgill. 1996. Diagnosis of *Legionella pneumophila*, *Mycoplasma pneumonia*, or *Chlamydia pneumoniae* lower respiratory infection using the polymerase chain reaction on a single throat swab specimen. *Diagn. Microbiol. Infect. Dis.* 24:7-14.

Ratcliff, R. M., S. C. Donnellan, J.A. Lanser, P.A. Manning, and M.W. Heuzenroeder. 1997. Interspecies sequence differences in the Mip protein from the genus *Legionella*: implications for function and evolutionary relatedness. *Mol. Microbiol.* 25:1149-1158.

Ratcliff, R. M., J. A. Lanser, P.A. Manning, and M.W. Heuzenroeder. 1998. Sequence-based classification scheme for the genus *Legionella* targeting the *mip* gene. *J. Clin. Microbiol.* 36:1560-1567.

Reeves, M. W., L. Pine, S.H. Hutner, J.R. George, and W.K. Harrell. 1981. Metal requirements of *Legionella pneumophila. J. Clin. Microbiol.* 13:688-695.

Reysenbach, A., L. J. Giver, G.S. Wickham, and N.R. Pace. 1992. Differential amplifcation of rRNA genes by polymerase chain reaction. *Appl. Environ. Microbiol.* 58:3417-3418.

Riffard, S., F. Lo Presti, P. Normand, F. Forey, M. Reyrolle, J. Etienne, and F. Vandenesch. 1998. Species identification of *Legionella* via intergenic 16S-23S ribosomal spacer PCR analysis. *Int. J. Syst. Bacteriol.* 48:723-730.

Rochelle, P. A., R. De Leon, A. Johnson, M. H. Stewart, and R. L. Wolfe. 1999. Evaluation of immunomagnetic separation for recovery of infectious *Cryptosporidium parvum* oocyts from environmental samples. *Appl. Environ. Microbiol.* 65:841-845.

Rogers, J., A. B. Dowsett, P. J. Dennis, J. V. Lee, and C. W. Keevil. 1994a. Influence of plumbing materials on biofilm formation and growth of *Legionella pneumophila* in potable water systems. *Appl. Environ. Microbiol.* 60:1842-1851.

Rogers, J., A. B. Dowsett, P.J. Dennis, J.V. Lee, and C.W. Keevil. 1994b. Influence of temperature and plumbing material selection on biofilm formation and growth of *Legionella pneumophila* in a model potable water system containing complex microbial flora. *Appl. Environ. Microbiol.* 60:1585-1592.

Rogers, J., A. B. Dowsett, J. V. Lee and C. W. Keevil. 1991. Chemostat studies of biofilm development on plumbing materials and the incorporation of *Legionella pneumophila*. In *Biodeterioration and Biodegradation*. Edited by H. W. Rossmore. Barking, United Kingdom: Elsevier Science Publishers, Ltd. 8:458-460.

Rogers, J., and C. W. Keevil. 1992. Immunogold and fluorescein immunolabelling of *Legionella pneumophila* within an aquatic biofilm visualized by using episcopic differential interference contrast microscopy. *Appl. Environ. Microbiol.* 58:2326-2330.

Rohr, U., M. Senger, and F. Selenka. 1996. Effect of silver and copper ions on the survival of *Legionella pneumophila* in tap water. *Zbl. Hyg.* 198:514-521.

Rohr, U., M. Senger, F. Selenka, R. Turley, and M. Wilhelm. 1999. Four years of experience with silver-copper ionization control of *Legionella* in a German university hospital hot water plumbing system. *Clin. Infect.* 29:1507-1511.

Rohr, U., S. Weber, R. Michel, F. Selenka, and M. Wilhelm. 1998. Comparison of free-living amoebae in hot water systems of hospitals with isolates for moist sanitary areas by identifying genera and determining temperature tolerance. *Appl. Environ. Microbiol.* 64:1822-1824.

Rooney-Varga, J. N., R. T. Anderson, J. L. Fraga,, D. Ringelberg, and D. R. Lorley. 1999. Microbial communities associated with anaerobic benzene degradation in a petroleum-contaminated aquifer. *Appl. Environ. Microbiol.* 65:3056-3063.

Rowbotham, T. J. 1980. Preliminary report on the pathogenicity of *Legionella pneumophila* for fresh water and soil amoebae. *J. Clin. Pathol.* 33:1179-1183.

Rowbotham, T. J. 1986. Current views on the relationships between *amoebae*, legionellae and man. *Isr. J. Med. Sci.* 22:678-689.

Rowbotham, T. J. 1989. Relations of legionellae with other microorganisms. Presented at Meeting on Legionellosis: Epidemiology, Prevention and Control, Geneva, Switzerland. 27-29 November 1989.

Rowbotham, T. J. 1993. *Legionella*-like amoebal pathogens. In *Legionella: Current Status and Emerging Perspectives*. Edited by J. M. Barbaree, R. F. Breiman, and A. P. Dufour. Washington, D.C.: American Society for Microbiology. 137-140.

Sanden, G. N., W. E. Morrill, B.S. Fields, R.F. Breiman, and J.M. Barbaree. 1992. Incubation of water samples containing amoebae improves detection of legionellae by the culture method. *Appl. Environ. Microbiol.* 58:2001-2004.

Saunders, N. A., N. Doshi, and T. G. Harrison. 1992. A second serogroup of *Legionella erythra* serologically indistinguishable from *Legionella rubrilucens*. *J. Appl. Bacteriol.* 72: 262-265.

Schofield, G. M., and R. Locci. 1985. Colonisation of components of a model hot water system by *Legionella pneumophila*. *J. Appl. Bacteriol.* 58:151-162.

Schwartz, T., S. Hoffman, and U. Obst. Formation and bacterial composition of young, natural biofilms obtained from public bank-filtered drinking water systems. *Water Research,* 32(9):2787-2797.

Segal, G., and H. A. Shuman. 1999. Intracellular multiplication of *Legionella pneumophila* in human host and environmental hosts. In *Microbial Ecology and Infectious Disease*. Edited by E. Rosenberg. Washington, D.C.: ASM Press. 170-186.

Shelton, B. G., W. D. Flanders, and G.K. Morris. 1994. Legionnaires' disease outbreaks and cooling towers with amplified *Legionella* concentrations. *Curr. Microbiol.* 28:359-363.

Smith-Somerville, H. E., V. Butz Huryn, C. Walker, and A.L. Winters. 1991. Survival of *Legionella pneumophila* in the cold water ciliate *Tetrahymena vorax*. *Appl. Environ. Microbiol.* 57:2742-2749.

Spino, D. F., E. W. Rice, and E. E. Geldreich. 1984. Occurrence of *Legionella* spp. and other aquatic bacteria in chemically contaminated groundwater treated by aeration. In *Legionella: Proceeding of the 2nd International Symposium.* Edited by C. Thornsberry, A. Balows, J. C. Feeley, and W. Jakubowski. Washington, D.C.: American Society for Microbiology. 318-320.

Stackebrandt, E., and B. M. Goebel. 1994. Taxonomic note: a place for DNA-DNA reassociation and 16S rRNA sequence analysis in the present species definition in bacteriology. *Int. J. Syst. Bacteriol.* 44:846-849.

Staley, J. T., and A. Konopka. 1985. Measurement of in situ activities of non photosynthetic microorganisms in aquatic and terrestrial habitats. *Ann. Rev. Microbiol.* 39: 321-346.

Starnbach, M. N., S. Falkow, and L.S. Tompkins. 1989. Species-specific detection of *Legionella pneumophila* in water by DNA amplification and hybridization. *J. Clin. Microbiol.* 27:1257-1261.

States, S. J., L. F. Conley, J.M. Kuchta, B.M. Oleck, M.J. Lipovich, R.S. Wolford, R.M. Wadowsky, A.M. McNamara, J.L. Sykora, and G. Keleti. 1987. Survival and multiplication of *Legionella pneumophila* in municipal drinking water systems. *Appl. Environ. Microbiol.* 53:979-986.

States, S. J., J. A. Podorski, L. F. Conley, W. P. Young, R. M. Wadawsky, J. N. Dowling, J. M. Kuchta, J. S. Navratil, and R. B. Yee. 1993. Temperature and the survival and multiplication of *Legionella pneumophila* associated with *Hartmannella vermiformis*. In *Legionella: Current Status and Emerging Perspectives*. Edited by J. M. Barbaree, R. F. Breiman, and A. P. Dufour. Washington, D.C.: American Society for Microbiology. 147-149.

Steele, T. W. 1993. Interactions between soil amoebae and soil Legionellae. In *Legionella: Current Status and Emerging Perspectives*. Edited by J. M. Barbaree, R. F. Breiman, and A. P. Dufour. Washington, D.C.: American Society for Microbiology. 140-142.

Steele, T. W., and A. M. McLennan. 1996. Infection of *Tetrahymena pyriformis* by *Legionella longbeachae* and other *Legionella* species found in potting mixes. *Appl. Environ. Microbiol.* 62:1081-1083.

Steele, T. W., C. V. Moore, and N. Sangster. 1990. Distribution of *Legionella longbeachae* serogroup 1 and other Legionellae in potting soils in Australia. *Appl. Environ. Microbiol.* 56(10):2984-2988.

Steinert, M., L. Emödy, R. Amann, and J. Hacker. 1997. Resuscitation of viable but nonculturable *Legionella pneumophila* Philadelphia JR32 by *Acanthamoeba castellanii*. *J. Clin. Microbiol.* 63:2047-2053.

Stout, J. E., M. G. Best, and V. L. Yu. 1986. Susceptibility of members of the family Legionellaceae to thermal stress: Implications for heat eradication methods in water distribution systems. *Appl. Environ. Microbiol.* 52:396-399.

Stout, J. E., M. G. Best, V.L. Yu, and J.D. Rihs. 1986. A note on symbiosis of *Legionella pneumophila* and *Tatlockia micdadei* with human respiratory flora. *J. Appl. Bacteriol.* 60:297-299.

Stout, J. E., V. L. Yu, and M.G. Best. 1985. Ecology of *Legionella pneumophila* within water distribution systems. *Appl. Environ. Microbiol.* 49:221-228.

Stout, J. E., V. L. Yu, and P. Muraca. 1985. Isolation of *Legionella pneumophila* from the cold water of hospital ice machines: Implications for origin and transmission of the organism. *Infect. Control* 6:141-146.

Stout, J. E., V. L. Yu, P. Muraca, J. Joly, N. Troup, and L. S. Tompkins. 1992. Potable water as a cause of sporadic cases of community-acquired Legionnaires' disease. *N. Engl. J. Med.* 326:151-155.

Struelens, M. J., N. Maes, F. Rost, A. Deplano, F. Jacobs, C. Liesnard, N. Bornstein, F. Grimont, S. Lauwers, and M. P. McIntyre, and E. Serruys. 1992. Genotypic and phenotypic methods for the investigation of a nosocomial *Legionella pneumophila* outbreak and efficacy of control measures. *J. Infect. Dis.* 166:22-30.

Suzuki, M. T., and S. J. Giovannoni. 1996. Bias caused by template annealing in the amplification of mixtures of 16S rRNA genes by PCR. *Appl. Environ. Microbiol.* 62:625-630.

Tang, P., and C. Krishnan. 1993. Legionellosis in Ontario, Canada: laboratory aspects. In *Legionella: Current Status and Emerging Perspectives*. Edited by J. M. Barbaree, R. F. Breiman, and A. P. Dufour. Washington, D.C.: American Society for Microbiology. 16-17.

Tang, P. W., S. Toma, and L.G. MacMillan. 1985. *Legionella oakridgensis*: Laboratory diagnosis of a human infection. *J. Clin. Microbiol.* 21(3):462-463.

Tang, P. W., S. Toma, C. W. Moss, A. G. Steigerwalt, T. G. Cooligan, and D. J. Brenner. 1984. *Legionella bozemanii* serogroup 2: a new etiological agent. *J. Clin. Microbiol.* 19:30-33.

Thacker, W. L., R. F. Benson, L. Hawes, H. Gidding, B. Dwyer, W. R. Mayberry, and D. J. Brenner. 1991. *Legionella fairfieldensis* sp. nov. isolated from cooling tower waters in Australia. *J. Clin. Microbiol.* 29:475-478.

Thacker, W. L., R. F. Benson, L. Hawes, W. R. Mayberry, and D. J. Brenner. 1990. Characterization of a *Legionella anisa* strain isolated from a patient with pneumonia. *J. Clin. Microbiol.* 28:122-123.

Thacker, W. L., R. F. Benson, R. B. Schifman, E. Pugh, A. G. Steigerwalt, W. R. Mayberry, D. J. Brenner, and H. W. Wilkinson. 1989. *Legionella tucsonensis* sp. nov. isolated from a renal transplant recipient. *J. Clin. Microbiol.* 27:1831-1834.

Thacker, W. L., R. F. Benson, J. L. Staneck, S. R. Vincent, W. R. Mayberry, D. J. Brenner, and H. W. Wilkinson. 1988. *Legionella cincinnatiensis* sp. nov. isolated from a patient with pneumonia. *J. Clin. Microbiol.* 26:418-420.

Thacker, W. L., R. F. Benson, H. W. Wilkinson, N. M. Ampel, E. J. Wing, A. G. Steigerwalt, and D. J. Brenner. 1986. 11th serogroup of *Legionella pneumophila* isolated from a patient with fatal pneumonia. *J. Clin. Microbiol.* 23:1146-1147.

Thacker, W. L., J. W. Dyke, R. F. Benson, D. H. Havlichek Jr, B. Robinson-Dunn, H. Stiefel, W. Schneider, C. W. Moss, W. R. Mayberry, and D. J. Brenner. 1992. *Legionella lansigensis* sp. nov. isolated from a patient with pneumonia and underlying chronic lymphocytic leukemia. *J. Clin. Microbiol.* 30:2398-2401.

Thacker, W. L., H. W. Wilkinson, R. F. Benson, and D. J. Brenner. 1987. *Legionella pneumophila* serogroup 12 isolated from human and environmental sources. *J. Clin. Microbiol.* 25:569-570.

Thacker, W. L., H. W. Wilkinson, R.F. Benson, S.C. Edberg, and D.J. Brenner. 1988. *Legionella jordanis* isolated from a patient with fatal pneumonia. *J. Clin. Microbiol.* 26:1400-1401.

Thacker, W. L., H. W. Wilkinson, B. B. Plikaytis, A. G. Steigerwalt, W. R. Mayberry, C. W. Moss, and D. J. Brenner. 1985. Second serogroup of *Legionella feeleii* strains isolated from humans. *J. Clin. Microbiol.* 22:1-4.

Thioulouse J., D. Chessel, S. Dolédec, and J.M. Olivier. 1997. ADE-4: A multivariate analysis and graphical display software. *Statistics and Computing* 7: 75-83.

Thompson, J. D., D. G. Higgins, and T.J. Gibson.. 1994. CLUSTAL W: Improving the sensitivity of progressive multiple sequence alignment through sequence weighting, positions-specific gaps penalties and weight matrix choice. *Nucleic Acids Research* 22:4673-4680.

Tison, D. L. 1987. Microbial ecology of *Legionella*. *J. Infect. Dis.* 156(5):852.

Tison, D. L., J. A. Baross, and R. J. Seidler. 1983. *Legionella* in aquatic habitats in the mount Saint Helens blast zone. *Curr. Microbiol.* 9:345-348.

Tison, D. L. and R. J. Seidler 1983. *Legionella* incidence and density in potable drinking water supplies. *Appl. Environ. Microbiol.* 45:337-339.

Tompkins, L. S., and J. S. Loutit. 1993. Detection of *Legionella* by molecular methods. In *Legionella: Current Status and Emerging Perspectives*. Edited by J. M. Barbaree, R. F. Breiman, and A. P. Dufour. Washington, D.C.: American Society for Microbiology. 163-168.

Towns, M. L., D. Fisher, and J. Moore. 1994. Community-acquired pneumonia due to *Legionella gormanii*. *Clin. Infect. Dis.* 18:265-266.

Toze, S., L. I. Sly, I. C. Macrae, and J. A. Fuerst. 1990. Inhibition of growth of *Legionella* species by heterotrophic plate count bacteria isolated from chlorinated drinking water. *Curr. Microbiol.* 21:139-143.

U.S. Geological Survey (USGS). 1999. *Sustainability of Ground Water Resources*. Circular #1186. Denver, Colo.: USGS.

Uyttendaele, M., I. Van Hoorde, and J. Debevere. 2000. The use of immuno-magnetic separation (IMS) as a tool in a sample preparation method for direct detection of *L. monocytogenes* in cheese. *Int. J. Food Microbiol.* 54:205-212.

Verma, V. K., D. J. Brenner, W. L. Thacker, R. F. Benson, G. Vesey, J. B. Kurtz, P. J. Dennis, A. G. Steigerwalt, J. S. Robinson, and C. W. Moss. 1992. *Legionella shakespearei* sp. nov., isolated from cooling tower water. *Int. J. Syst. Bacteriol.* 42:404-407.

Visca, P., P. Goldoni, P. C. Luck, J. H. Helbig, L. Cattani, G. Giltri, S. Bramati, and M. Castellani-Pastoris. 1999. Multiple types of *Legionella pneumophila* serogroup 6 in a hospital heated-water system associated with sporadic infections. *J. Clin. Microbiol.* 37:2189-2196.

Wadowsky, R. M., T. M. Wilson, N.J. Kapp, A.J. West, J.M Kuchta, S.J. States, J.N. Dowling, and and R.B. Yee. 1991. Multiplication of *Legionella* spp. in tap water containing *Hartmannella vermiformis*. *Appl. Environ. Microbiol.* 57:1950-1955.

Wadowsky, R. M., R. Wolford, A.M. McNamara, and R.B. Yee. 1985. Effect of temperature, pH, and oxygene level on the multiplication of naturally occuring *L. pneumophila* in potable water. *Appl. Environ. Microbiol.* 49:1197-1205.

Wadowsky, R. M., and R. B. Yee. 1981. Glycine-containing selective medium for isolation of Legionellaceae from environmental specimens. *Appl. Environ. Microbiol.* 42(5):768-772.

Wadowsky, R. M., and R. B. Yee. 1983. Satellite growth of *Legionella pneumophila* with an environmental isolate of *Flavobacterium breve*. *Appl. Environ. Microbiol.* 46:1447-1449.

Wadowsky, R. M., and R. B. Yee. 1985. Effect of non-Legionellaceae bacteria on the multiplication of *Legionella pneumophila* in potable water. *Appl. Environ. Microbiol.* 49(5):1206-1210.

Wang, G. C.-Y., and Y. Wang. 1997. Frequency of formation of chimeric molecules as a consequence of PCR coamplification of 16S rRNA genes from mixed bacterial genomes. *Appl. Environ. Microbiol.* 63:4645-4650.

Watts, J. C., M. D. Hicklin, B. M. Thomason, C. S. Callaway, and A. J. Levine. 1980. Fatal pneumonia caused by *Legionella pneumophila*, serogroup 3: demonstration of the bacilli in extrathoracic organs. *Ann. Internal Med.* 92:186-188.

Whitmore, T. N., and S. Sidorowicz. 1995. Novel methods for the concentration and selective purification of microorganisms and their potential application for water analysis. *Microbiol. Europe* 3:16-22.

Wilkinson, H. W., V. Drasar, W.L. Thacker, R.F. Benson, J. Schindler, B. Potuznikova, W.R. Mayberry, and D.J. Brenner. 1988. *Legionella moravica* sp. nov. and *Legionella brunensis* sp. nov. isolated from cooling-tower water. *Ann. Inst. Pasteur/Microbiol.* 139:393-402.

Wilkinson, H. W., W. L. Thacker, R. F. Benson, S. S. Polt, E. Brookings, W. R. Mayberry, D. J. Brenner, R. G. Gilley, and J. K. Kirklin. 1987. *Legionella birminghamensis* sp. nov. isolated from a cardiac transplant recipient. *J. Clin. Microbiol.* 25:2120-2122.

Wilkinson, H. W., W. L. Thacker, A. G. Steigerwalt, D. J. Brenner, N. M. Ampel, and E. J. Wing. 1985. Second serogroup of *Legionella hackeliae* isolated from a patient with pneumonia. *J. Clin. Microbiol.* 22:488-489.

Wilson, I. G. 1997. Inhibition and facilitation of nucleic acid amplification. *Appl. Environ. Microbiol.* 63:3741-3751.

Winn, W. C. 1988. Legionnaires' disease : Historical perspective. *Clin. Microbiol. Rev.* 1: 60-81.

Winn, W. C. 1995. *Legionella.* In *Manual of Clinical Microbiology*, 6th ed. Edited by P. R. Murray, E. J. Baron, M. A. Pfaller, F. C. Tenover, and R. H. Yolken. Washington, D.C.: ASM Press. 533-544.

Yamamoto, H., Y. Hashimoto, and T. Ezaki. 1993. Comparison of detection methods for *Legionella* species in environmental water by colony isolation, fluorescent antibody staining, and polymerase chain reaction. *Microbiol. Immunol.* 37:617-622.

Yamamoto, H., Y. Hashimoto, and E. Takayuki. 1996. Study of nonculturable *Legionella pneumophila* cells during multiple-nutrient starvation. *FEMS Microbiol. Ecol.* 20:149-154.

Yates, M. V. 1997. Placement and protection of drinking water wells. In *Manual of Environmental Microbiology.* Edited by C. J. Hurst, G. R. Knudsen, M. J. McInerney, L. D. Stetzenbach, and M. V. Walter. Washington, D.C.: American Society for Microbiology. 608-616.

Yee, R. B., and R. M. Wadowsky. 1982. Multiplication of *Legionella pneumophila* in unsterilized tap water. *Appl. Environ. Microbiol.* 43:1330-1334.

Zacheus, O. M., and P. J. Martikainen. 1994. Occurrence of legionellae in hot water distribution systems of Finnish apartment buidings. *Can. J. Microbiol.* 40:993-999.

Zacheus, O. M., and P. J. Martikainen. 1996. Effect of heat flushing on the concentrations of *Legionella pneumophila* and other heterotrophic microbes in hot water systems of apartment buildings. *Can. J. Microbiol.* 42:811-818.

ABBREVIATIONS

AOC	assimilable organic carbon
AP	samples or plates from samples that received acid treatment
ATCC	American Type Culture Collection
ATP	adenosine triphosphate
AwwaRF	Awwa Research Foundation
AWWSCo	American Water Works Service Company
BART™	Biological Activity Reaction Test
CDC	Centers for Disease Control and Prevention
CFU	colony forming unit
CREM	Centre for Research on Environmental Microbiology
CTP	cytidine triphosphate
CWS	community water system
DFA	direct fluorescent antibody
DGGE	denaturing gradient gel electrophoresis
DNA	deoxyribonucleic acid
DP	samples plated directly without other treatments, or plates from same
fg	femtogram – unit of weight
FISH	fluorescent in-situ hybridization
GTP	guanosine triphosphate
GWR	Groundwater Rule
h	hour
HT	samples that were heat treated before plating, or the resulting plates
IMS	immunomagnetic separation
IMS-PCR	immunomagnetic separation combined with polymerase chain reaction
L	liter
LD	legionnaires' disease
LLAP	*Legionella*-like amoebal pathogen
Mab	monoclonal antibody
MCLG	maximum contaminant level goal
mg	milligram - unit of weight
Mip	macrophage infectivity potentiator protein
mip	gene encoding for Mip
mL	millilitre – unit of volume
mM	millimolar – measure of molarity

μg	microgram - unit of weight
μL	microliter – unit of volume
μmho	micromho – unit of electrical conductivity – equivalent to μS
μS	micro Siemens – unit of electrical conductivity – equivalent to μmhos
NCTC	National Collection Type Culture
NTNCWS	non-transient non-community water system
NTU	nephelometric turbidity unit
PAC	Project Advisory Committee
PCM	Polish Culture Collection
PCR	polymerase chain reaction
pH	measure of acidity or alkalinity
PLFA	phospholipin fatty acid analysis
ppm	parts per million
PVC	polyvinyl chloride
QA/QC	quality assurance/quality control
RNA	ribonucleic acid
spp.	multiple species
SWTR	Surface Water Treatment Rule
TBE	tris borate EDTA solution
THMs	trihalomethanes
TNCWS	transient non-community water system
TOC	total organic carbon
TTP	thymidine triphosphate
U.S. EPA	U.S. Environmental Protection Agency
UV	ultraviolet
w/v	weight per volume
WWW	World Wide Web

IWA Publishing's authorised EU representative for General Product Safety Regulations is Diane D'Arras, 15 rue Duret, 75116 Paris, France, e-mail: safety@iwap.co.uk.

Printed and bound by CPI Group (UK) Ltd, Croydon, CR0 4YY
03/10/2026
02227463-0003